Advances in Intelligent Systems and Computing

Volume 911

Series editor

Janusz Kacprzyk, Systems Research Institute, Polish Academy of Sciences,
Warsaw, Poland
e-mail: kacprzyk@ibspan.waw.pl

The series "Advances in Intelligent Systems and Computing" contains publications on theory, applications, and design methods of Intelligent Systems and Intelligent Computing. Virtually all disciplines such as engineering, natural sciences, computer and information science, ICT, economics, business, e-commerce, environment, healthcare, life science are covered. The list of topics spans all the areas of modern intelligent systems and computing such as: computational intelligence, soft computing including neural networks, fuzzy systems, evolutionary computing and the fusion of these paradigms, social intelligence, ambient intelligence, computational neuroscience, artificial life, virtual worlds and society, cognitive science and systems, Perception and Vision, DNA and immune based systems, self-organizing and adaptive systems, e-Learning and teaching, human-centered and human-centric computing, recommender systems, intelligent control, robotics and mechatronics including human-machine teaming, knowledge-based paradigms, learning paradigms, machine ethics, intelligent data analysis, knowledge management, intelligent agents, intelligent decision making and support, intelligent network security, trust management, interactive entertainment, Web intelligence and multimedia.

The publications within "Advances in Intelligent Systems and Computing" are primarily proceedings of important conferences, symposia and congresses. They cover significant recent developments in the field, both of a foundational and applicable character. An important characteristic feature of the series is the short publication time and world-wide distribution. This permits a rapid and broad dissemination of research results.

More information about this series at http://www.springer.com/series/11156

Mostafa Ezziyyani
Editor

Advanced Intelligent Systems for Sustainable Development (AI2SD'2018)

Vol 1: Advanced Intelligent Systems Applied to Agriculture

Editor
Mostafa Ezziyyani
Computer Sciences Department,
Faculty of Sciences and Techniques
of Tangier
Abdelmalek Essaâdi University
Souani Tangier, Morocco

ISSN 2194-5357 ISSN 2194-5365 (electronic)
Advances in Intelligent Systems and Computing
ISBN 978-3-030-11877-8 ISBN 978-3-030-11878-5 (eBook)
https://doi.org/10.1007/978-3-030-11878-5

Library of Congress Control Number: 2019930141

This Springer imprint is published by the registered company Springer Nature Switzerland AG
The registered company address is: Gewerbestrasse 11, 6330 Cham, Switzerland

Preface

Overview

The purpose of this volume is to honour myself and all colleagues around the world that we have been able to collaborate closely for extensive research contributions which have enriched the field of Applied Computer Science. Applied Computer Science presents a appropriate research approach for developing a high-level skill that will encourage various researchers with relevant topics from a variety of disciplines, encourage their natural creativity, and prepare them for independent research projects. We think this volume is a testament to the benefits and future possibilities of this kind of collaboration, the framework for which has been put in place.

About the Editor

Prof. Dr. **Mostafa Ezziyyani,** IEEE and ASTF Member, received the "Licence en Informatique" degree, the "Diplôme de Cycle Supérieur en Informatique" degree and the PhD "Doctorat (1)" degree in Information System Engineering, respectively, in 1994, 1996 and 1999, from Mohammed V University in Rabat, Morocco. Also, he received the second PhD degree "Doctorat (2)" in 2006, from Abdelmalek Essaadi University in Distributed Systems and Web Technologies. In 2008, he received a Researcher Professor **Ability Grade. In 2015, he receives a PES grade —the highest degree at Morocco University.** Now he is a Professor of Computer Engineering and Information System in Faculty of Science and Technologies of Abdelmalek Essaadi University since 1996.

His research activities focus on the modelling databases and integration of heterogeneous and distributed systems (with the various developments to the big data, data sciences, data analytics, system decision support, knowledge management, object DB, active DB, multi-system agents, distributed systems and mediation). This research is at the crossroads of databases, artificial intelligence, software engineering and programming.

Professor at Computer Science Department, Member MA laboratory and responsible of the research direction Information Systems and Technologies, he formed a research team that works around this theme and more particularly in the area of integration of heterogeneous systems and decision support systems using WSN as technology for communication.

He received the first WSIS prize 2018 for the Category C7: ICT applications: E-environment, First prize: MtG—ICC in the regional contest IEEE - London UK Project: "World Talk", The qualification to the final (Teachers-Researchers Category): Business Plan Challenger 2015, EVARECH UAE Morocco. Project: «Lavabo Intégré avec Robinet à Circuit Intelligent pour la préservation de l'eau», First prize: Intel Business, Challenge Middle East and North Africa—IBC-MENA. Project: «Système Intelligent Préventif Pour le Contrôle et le Suivie en temps réel des Plantes Médicinale En cours de Croissance (PCS: Plants Control System)», Best Paper: International Conference on Software Engineering and New Technologies ICSENT'2012, Hammamat-Tunis. Paper: «Disaster Emergency System Application Case Study: Flood Disaster».

He has authored three patents: (1) device and learning process of orchestra conducting (e-Orchestra), (2) built-in washbasin with intelligent circuit tap for water preservation. (LIRCI) (3) Device and method for assisting the driving of vehicles for individuals with hearing loss.

He is the editor and coordinator of several projects with Ministry of Higher Education and Scientific Research and others as international project; he has been involved in several collaborative research projects in the context of ERANETMED3/PRIMA/H2020/FP7 framework programmes including project management activities in the topic modelling of distributed information systems reseed to environment, Health, energy and agriculture. The first project aims to propose an adaptive system for flood evacuation. This system gives the best decisions to be taken in this emergency situation to minimize damages. The second project aims to develop a research dynamic process of the itinerary in an events graph for blind and partially signet users. Moreover, he has been the principal investigator and the project manager for several research projects dealing with several topics concerned with his research interests mentioned above.

He was an invited professor for several countries in the world (France, Spain Belgium, Holland, USA and Tunisia). He is member of USA-J1 programme for TCI Morocco Delegation in 2007. He creates strong collaborations with research centres in databases and telecommunications for students' exchange: LIP6, Valencia, Poitier, Boston, Houston, China.

He is the author of more than 100 papers which appeared in refereed specialized journals and symposia. He was also the editor of the book "New Trends in Biomedical Engineering", AEU Publications, 2004. He was a member of the Organizing and the Scientific Committees of several symposia and conferences dealing with topics related to computer sciences, distributed databases and web technology. He has been actively involved in the research community by serving as reviewer for technical, and as an organizer/co-organizer of numerous international and national conferences and workshops. In addition, he served as a programme committee member for international conferences and workshops.

He was responsible for the formation cycle "Maîtrise de Génie Informatique" in the Faculty of Sciences and Technologies in Tangier since 2006. He is responsible too and coordinator of Tow Master "DCESS - Systèmes Informatique pour Management des Entreprise" and "DCESS - Systèmes Informatique pour Management des Enterprise". He is the coordinator of the computer science modules and responsible for the graduation projects and external relations of the Engineers Cycle "Statistique et Informatique Décisionnelle" in Mathematics Department of the Faculty of Sciences and Technologies in Tangier since 2007. He participates also in the Telecommunications Systems DESA/Masters, "Bio-Informatique" Masters and "Qualité des logiciels" Masters in the Faculty of Science in Tetuan since 2002.

He is also the founder and the current chair of the blinds and partially signet people association. His activity interests focus mainly on the software to help the blinds and partially signet people to use the ICT, specifically in Arabic countries. He is the founder of the private centre of training and education in advanced technologies AC-ETAT, in Tangier since 2000.

Mostafa Ezziyyani

Contents

Android Applications Analysis Using PerUpSecure

Latifa Er-rajy(✉), My Ahemed El Kiram, and Mohamed El Ghazouani

Department of Computer Sciences, University Cadi Ayyad, Marrakesh, Morocco
errajy.latifa@gmail.com, kiram@uca.ac.ma,
Mohamed.elghazouani63@gmail.com

Abstract. Since its introduction in 2008, Google's Android has been a blazing success, far outstripping the market share of all other mobile operating systems. Android ships more than one billion new devices each year, and more than 1.5 million new devices are activated every day. This growth was not without pain, however. Recent measures estimate that 96–97% of today's mobile malware targets the Android operating system, and 73% of them are specifically designed to satisfy profit motives. In addition, as the system becomes more popular and scrutinized, the number of vulnerabilities identified has exploded. The security of Android is a key issue for Google, the mobile OS is - by far - the most popular in the world. The Android security mechanism is founded on an instrument that gives users all the information about the permissions requested by the application before installing it. The main benefit of this Android permission system is to provide users an overview of the application by showing them the requested permissions list, which can help raise awareness of its risks on their private data. However, we still do not have enough information to allow us to say that standard users are able to clearly understand the permissions requested and their implications for their security. In this article, we present a tool called "PerUpSecure" multiphases that combines dynamic and static analysis and contrary to what we know about the installation process of Android applications that puts in front of the user only two options, either he accepts all requested permissions or he cancels the installation, our proposed tool allows the user to install any application with only the necessary permissions. At the end of this article, we present the analysis results of a set of normal applications and malicious programs collected from different markets.

Keywords: Static analysis · Permissions · Sandboxing · Repackaging · Tool

1 Introduction

Today, Android captures about 85% of global smartphone shipments and continues to generate growth of 3.2% on average per year [1]. By 2021, the volume of terminals will climb over 1.5 billion euros units [2]. This popularity continues to grow not only among users, but also among mobile application developers who take advantage of the effect that Android is an open source OS. For all these reasons, Android has become a very interesting and very easy target for cyber-attacks [3]. Indeed, the Android

M. Ezziyyani (Ed.): AI2SD 2018, AISC 911, pp. 1–13, 2019.
https://doi.org/10.1007/978-3-030-11878-5_1

malwares number has grown up recently. And these attacks are no less dangerous than malware targeting desktop computers [4].

On 2017, Android malware already accounted for 99% of all mobile threats. Their expansion continued, with growth of 70% between January and August, according to the F-Secure Laboratory [5].

Android smartphones are protected by a permissions-based framework that limits third-party applications' access to sensitive resources such as the SMS database and external storage on Android smartphones [6].

One of the questions that can be asked about the permissions system is the existence (or not) of a coherent underlying theoretical model. Does each permission correspond to a set of unique and strictly delimited tasks? this is the same question that arises with the model of capabilities under Linux, and in this second case the answer is "no": CAP_SYS_MODULE allows for example to load arbitrary kernel modules, and thus to compromise completely the integrity of the kernel [7]. In the case of Android, we may suspect that the permission SYSTEMPROPERTIES_WRITE defined by the operator Sprint probably has a non-zero intersection with the permission WRITE_SETTINGS defined by Google, while probably unknown to the analysis tools [8].

The practical value of a mathematical study on Android permissions is however limited. Indeed, the two major risks [9, 10] encountered in nature are:

- Malicious applications that require excessively large permissions, and abuse the trust of users;
- Unpermitted malicious applications that exploit system vulnerabilities to gain access to the root account and completely bypass the security model.

Almost all existing applications require unlimited access to the network, very often to retrieve advertising content. In practice, this means that the risk posed by an application is not limited to a set of static permissions: for example, malicious content may be sent to the WebKit rendering engine at any time through the advertising channel, and thus cause control of a given phone through the application.

The requested permissions may change during an update [11]. Most users trust applications they have already installed since they may access to sensitive resources can lead to loss of money. For example, Android malware can send messages, make calls, and generate a large amount of network data without users being aware of it. Moreover, access to sensitive resources can lead to leakage of users' private information stored in their mobile devises such as contacts, emails and even credit card numbers which can have a dangerous impact on them.

In this paper, we present the test results of PerUpSecure tool that has three mean stages: static analysis, dynamic analysis and applications repackaging. after the user install PerUpSecure on his Android device, the tool inserts immediately instrumentation code into arbitrary Android applications and the monitoring code that intercepts an application's interactions with the system in case of updates in order to enforce various security policies by watching these updates and the permissions that may be added without the user permission.

2 PerUpSecure Tool

PerUpSecure is a tool sandbox based dynamic and static analysis that evaluate android applications permissions in installation time through several levels after using APKtool [12] which is integrated in our proposed tool in order to extract manifest file from APK file. Figure 1 presents some screenshots of PerUpSecure.

In addition, we used Java reflection [13] to get all API calls contained in manifest file, which will be used in static analysis phase. Generally, each application before being able to be installed, goes through two analysis levels: static and dynamic. The biggest Advantage of PerUpSecure tool is to enable the user to install an application with only the necessary permissions instead of accepting all the requested permissions or canceling completely the installation.

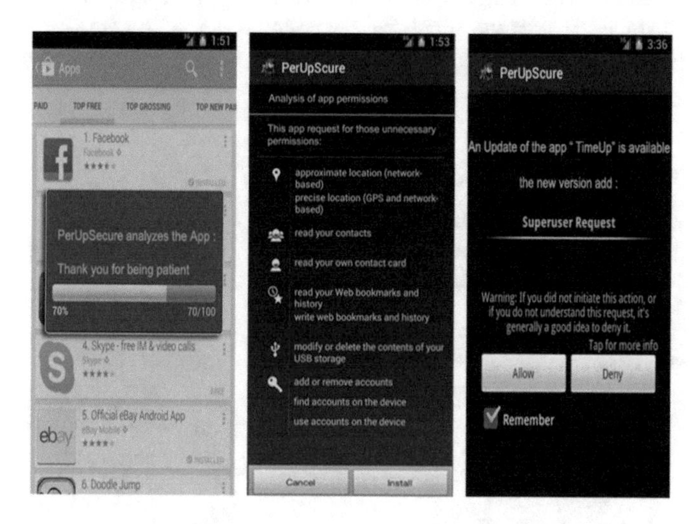

Fig. 1. PerUpSecure screenshots

2.1 How PerUpSecure Work

PerUpSecure tool has three major process as it is shown in Fig. 2: static analysis, dynamic analysis and applications repackaging. after the user install PerUpSecure on his Android device, the tool inserts immediately instrumentation code into arbitrary Android applications and the monitoring code that intercepts the application interactions with the system in case of updates to enforce various security policies by watching these updates and the permissions that may be added without the user permission.

Static Analysis: The first thing that the tool does during the static analysis phase is to scan the Android application package (APK) for special patterns (for example, Runtime.Exec ()), which is used to classify the application to facilitate and speed up the reading of the database. Our implementation of static analysis is run offline which makes it light enough to run on the Android device. However, dynamic analysis requires emulation on a more powerful machine.

When the user wants to install an application for the first time, PerUpSecure tool preforms a static analysis in which compares the permissions requested by the application with a predefined database that contains that all the permission requirements for every API calls and permission specifications for more than one version of Android to make a clear decision about what kind of permissions are requested by the application (necessary, unnecessary or dangerous). In fact, this database contains almost all API calls that can be required by Android applications.

Dynamic Analysis: This analysis is based on an Android virtual device founded on QEMU [14] similar to Android SDK.

In this work, the application is installed in the standard Android emulator from the Google Android SDK. Once application installation ended, Monkey Tool [15] which is a program installed inside the emulator generates a set of random pseudo-streams presented the user events such as clicks, keys, gestures, and a number of system-level events.

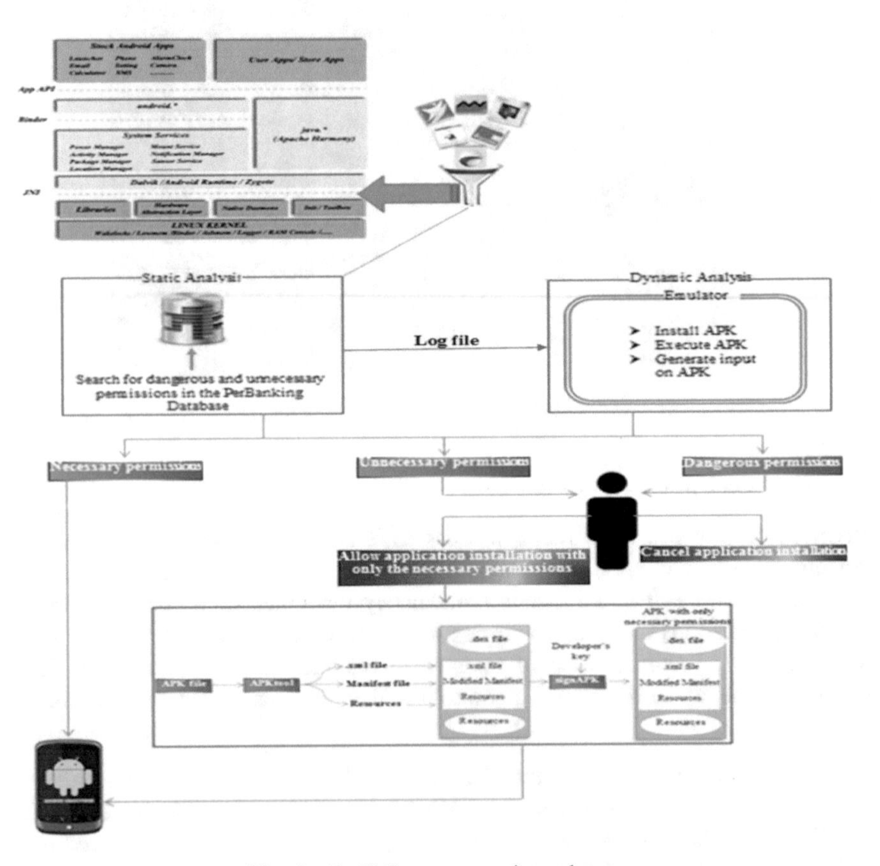

Fig. 2. PerUpSecure operation scheme

The Monkey was mainly invented for stress testing applications. Our tool is placed in the kernel space and shakes the system calls for logging. The dynamic analysis result helps us to record application behavior at the system level. The resulting log file is reduced to a mathematical vector to ensure a better analysis. The kernel module ensures that every occurrence of a system call is saved with the required permission. This ensures the registration of a complete system state and no malicious activity can be hidden. An application system call log is saved in a separate file.

At the end of the dynamic analysis, PerUpSecure Tool receives two logs files. It examines first the result of code analysis. If the application contains a malicious code, it installation is automatically abort and a popup window is shown to the user explains why the application installation is cancelled. Otherwise, PerUpSecure Tool examines the permission result, if there are unnecessary or dangerous permissions requested by the application, then, the tool send to user a notification that invites him either to allow application installation with only the necessary permission or cancel completely the installation. In the case where the user chooses to continue the installation, the tool repackages the application in order to delete the unnecessary permissions before allowing its installation.

Application Repackaging: In this process, we have to delete the unnecessary and dangerous permissions from the application manifest file.

Each application is going through five steps before being ready to be installed in to user device because the code in the APK file is so difficult to read by a human, since it contains Dalvik bytecode (Dex format): extraction, decoding, modifying, encoding, and packing. Figure 3 shows a rough overview of the process used to modify an existing .apk file. The purpose of the extraction and decoding steps is to transform an .apk file into an easily editable form. The modification step is an application specific step which involves reading and modifying the bytecode. The encoding and packing steps create a new .apk file from the modified files.

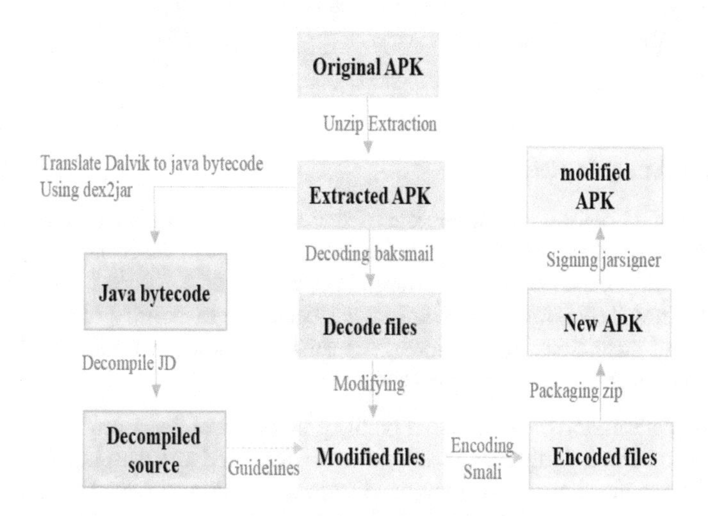

Fig. 3. Application repackaging steps

3 Experiments Results

Our dataset consists of 2279 Android apps. We made the application collection during January-February 2018. During this period, we explored PlayStore [16] and 7 third-party stores which are: AndroidBest [17], AndroidDrawer [18], AndroidLife [19] Anruan [20], AppsApk [21], PandaApp [22], and SlideME [23].

Table 1 lists all the markets used in the analysis. During this analysis, we discovered that all alternative markets give users the ability to download applications without authentication. So, they need simply to the right URL to download an application.

In exploring the web pages of the markets, we have discovered that for some of them, these URLs can be easily predicted. For example, all applications on the AndroidBest market can be downloaded simply by entering the following URLs "download.php?Id=n", where n varies from 0 to the order number of the last downloaded application. Also, in the case of AndroidDrawer, a crawler was developed to parse the html pages and extract the URLs used to download the applications. After downloading the applications, we performed a two-step analysis of our dataset using our "PerUpSecure" tool. Our dataset occupies 17.4 GB of disk space.

Table 1. Applications set markets

Market	Market link	Applications number
AndroidBest	http://androidbest.ru/	62
AndroidDrawer	http://www.androiddrawer.com/	157
AndroidLife	http://androidlife.ru/	678
Anruan	http://www.anruan.com/	232
AppsApk	http://www.appsapk.com/	79
Google PlayStore	https://play.google.com/store/apps	223
PandaApp	http://android.pandaapp.com/	543
SlideMe	http://slideme.org/	305
Total		2279

3.1 Static Analysis Results

As we mentioned before, the initial goal of static analysis is to calculate the maximum set of Android permissions that an application may need to function properly.

We randomly chose 400 applications from the set of 2279 that we have and we analyzed them with PerUpSecure. Our tool has been able to identify 118 applications as dangerous and suspicious applications because they require permissions that are not necessary and at the same time dangerous.

Dangerous Permissions: We focused on the prevalence of dangerous permissions. As we mentioned before, dangerous permissions are displayed as a warning to users during the applications installation and may have serious security ramifications on the user's personal data. We noted that 82% of the applications analyzed have at least one dangerous permission. Permissions in Android are grouped into feature categories. This provides a relative measure part of the protected API that is used by the applications. A small number of permissions are required very frequently. We find that 24% of applications request INTERNET as their only dangerous permission. we also found that 38% of applications combine between three dangerous permissions, as for example 16% of applications require the permissions: CONTACTS, SMS and PHONE at the same time, which means that these applications have the power to control and use the user mobile phone to call numbers and send SMS without the user's awareness what it will probably lead to dangerous consequences on his banking transactions.

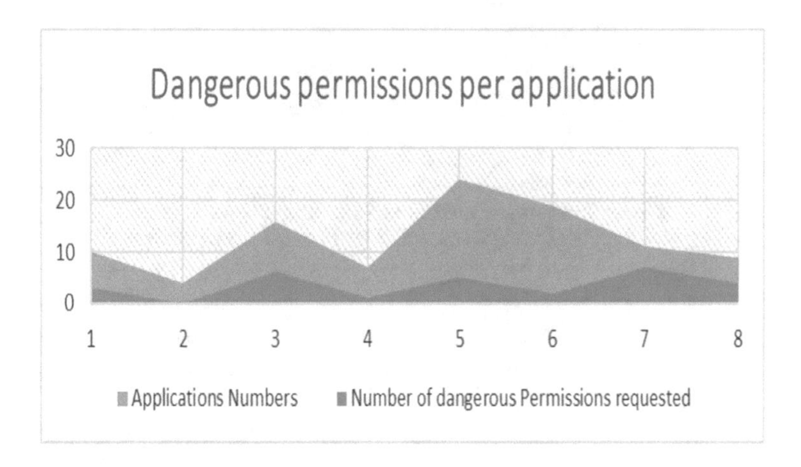

Fig. 4. Dangerous permission distribution

Although many applications ask for at least one dangerous permission, the total number of permission requests is typically low. The most highly privileged application in our set asks for less than half of the available 56 dangerous permissions. Figure 4 shows the distribution of dangerous permission requested.

Several important categories are requested relatively infrequently, which is a positive finding. Permissions in the PERSONAL_INFO and COST_MONEY categories are only requested by 5% of applications. The PERSONAL_INFO category includes permissions associated with the user's contacts, calendar, etc.; COST_MONEY permissions let applications send text messages or make phone calls without user confirmation.

Users have reason to be suspicious of applications that ask for permissions in these categories. As showed by Table 2, nearly all applications (82%) ask for at least one dangerous permission, which indicates that users frequently install applications with dangerous permissions. We were interested in the dangerous permissions most frequently requested by all the applications we analyzed. Figure 5 below illustrates the results of the analysis obtained.

Table 2. Applications with at least one dangerous permission in each category

Category	Applications %
NETWORK**	66%
SYSTEM_TOOLS	39.7%
STORAGE**	34.1%
LOCATION**	26%
PHONE_CALLS	35%
PERSONAL_INFO	13%
HARDWARE_CONTROLS	17%
COST_MONEY	9%
MESSAGES	5%
ACCOUNTS	2%
DEVELOPMENT_TOOLS	0%

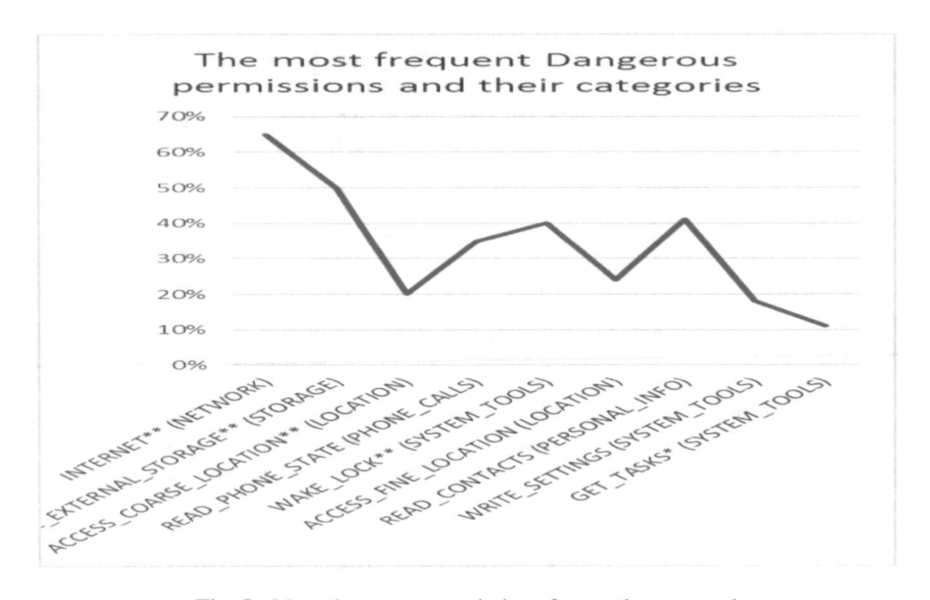

Fig. 5. Most dangerous permissions frequently requested

As we can notice the permissions: INTERNET (NETWORK), WRITE_ EXTERNAL_STORAGE (STORAGE), READ_PHONE_STATEMENT and WAKE_LOCK are the most frequent dangerous permissions requested.

WAKE _LOCK permission Allows to an application the using of PowerManager WakeLocks to keep processor from sleeping or screen from dimming. It is the same thing for the permission ACCESS_FINE_LOCATION that allow to an application to access precise location of the Mobile device owner.

3.2 Dynamic Analysis Results

To perform the dynamic analysis and test the good work of our virtual emulator, we downloaded from Google PlayStore about 600 free applications (the results in Table 3).

API Calls Results: As we mentioned before, our emulator can analyze API calls and count the number of methods and classes have permission checks. This number of checks for permissions, unused permissions, hierarchical permissions, permission granularity, and class characteristics is shown above.

Number of Permissions Checks: Our tool recognized 1447 API calls with permission control, among these calls: API 889 present methods of normal API classes, and 586 have RPC stubs methods exploited in communication with Android system services. In addition, the dynamic analysis results show that 28 API calls have permission checks in an additional part of the API added by a manufacturer, for a total of 1503 API calls that have permission.

Signature/System Permissions: We noticed that 14% of normal API calls and 38% of APIs with RPC stubs are protected with Signature/System permissions. The results show that some permissions are defined by the platform but never used in the API calls. For example, BRICK permission is never used, although it is cited as an example of a particularly dangerous permission [24]. The only use of BRICK permission is in the dead code which is not capable of causing any damage to the device. Dynamic analysis revealed that 25 of the 134 permissions set by Android are unused which makes them totally unnecessary. For each case where a permission was never found during the scan, we manually searched the source tree to verify that the permission is not being used. After running our virtual emulator on multiple devices, we discovered that one of the unused permissions is being used by the custom classes that HTC and Samsung have added to existing API calls to support 4G connection on their phones.

Hierarchical Permissions: Since the names of many permissions we can easily conclude that there are hierarchical relationships between these permissions. Intuitively, we expect stronger permissions to be substitutable for lower permissions for the same resource. However, we have no evidence of planned hierarchy. In the end, we concluded that BLUETOOTH_ADMIN cannot replace BLUETOOTH, and that WRITE_CONTACTS cannot replace READ_CONTACTS. Likewise, CHANGE_-WIFI_STATE cannot be used instead of ACCESS_WIFI_STATE. Then, ACCESS_-COARSE_LOCATION and ACCESS_FINE_LOCATION is the only permission pair that has a hierarchical relationship. So, any method that admits COARSE accepts FINE as an alternative. This is the only exception to this, which can be a bug.

Granularity of Permission: In the case where a single permission is made to a varied set of features, this applies that applications that request permission for a subset of the feature will have unnecessary access to the rest. The purpose of Google and specifically Android is to avoid this by dividing the feature into multiple permissions, then their approach has been shown to benefit from the more secure platform [25]. So, we looked at the division of the Bluetooth feature as a case study since the Bluetooth permissions

Table 3. Applications set analysis results

Category	Applications number	Permissions Requsted Number	Unnecessary Permissions Number	Number of DCL	DCL calls Number	JV Reflexion	#JV ref calls
APP WALLPAPER	18	48	24	12	96	306	**351**
APP WIDGETS	42	19	5	4	38	87	**390**
BOOKS AND REFERENCE	10	66	17	16	132	113	**369**
BUSINESS	50	48	4	12	96	138	**411**
COMICS	17	94	18	23	188	110	**363**
COMMUNICATION	12	25	8	6	50	97	**298**
EDUCATION	40	68	27	17	136	105	**305**
ENTERTAINMENT	7	34	13	8	68	70	**307**
FINANCE	19	38	7	9	76	63	**418**
GAME	20	22	10	5	44	103	**528**
HEALTH AND FITNESS	52	77	8	19	154	168	**532**
LIBRARIES AND DEMO	56	88	21	22	176	193	**532**
LIFESTYLE	51	104	19	26	208	132	**405**
MEDIA AND VIDEO	16	28	12	7	56	100	**394**
MEDICAL	58	64	15	16	128	90	**502**
MUSIC AND AUDIO	13	18	7	4	36	105	**551**
NEWS AND MAGAZINES	51	79	33	19	158	177	**575**
PERSONALIZATION	51	90	23	22	180	199	**561**
PHOTOGRAPHY	59	109	18	27	218	138	**590**
PRODUCTIVITY	16	27	19	6	54	105	**575**
SHOPPING	59	67	23	16	134	190	**695**
SOCIAL	52	119	22	29	238	212	**648**
SPORTS	50	97	24	24	194	193	**581**
TOOLS	50	90	42	22	180	154	**513**
TRANSPORTATION	16	58	28	14	116	118	**331**
TRAVEL AND LOCAL	46	50	30	12	100	125	**3367**
WEATHER	17	70	16	17	140	65	**3241**
Total:	**948**	**1697**	**493**	**414**	**3653**	**3653**	**18333**

are the most highly verified permissions. Our analysis shows that the two Bluetooth permissions are applied to 6 major classes. They are split between methods that change state (BLUETOOTH_ADMIN) and methods that get information about the device (BLUETOOTH). So, the BluetoothAdapter class is one of several that exploit Bluetooth permissions and appropriately divides most of its permissions assignments. Yet, this receives some inconsistencies.

Unnecessary Permission: comparing two files shows that 45% applications have unnecessary permissions, which are at the same time dangerous. Table 4 below shows that almost all unnecessary requested by applications in our set are dangerous.

Table 4. Unnecessary permissions requested

PERMISSION	%	PERMISSION LEVEL
ACCESS_NETWORK_STATE	25%	Normal
READ_PHONE_STATE	45%	Dangerous
ACCESS_WIFI_STATE	38%	Normal
WAKE_LOCK	5%	Dangerous
WRITE_EXTERNAL_STORAGE	27%	Dangerous
ACCESS_LOCATION	36%	Dangerous
PHONE	55%	Dangerous
SMS	65%	Dangerous
CAMERA	15%	Dangerous
INTERNET	85%	Dangerous
CONTACTS	30%	Dangerous
DEVICE ID & CALL INFORMATION	28%	Dangerous
PHOTOS/MEDIA/FILES	34%	Dangerous

We also found that 42.5% of the applications permissions are overcrowded, with a total of 39% unnecessary permissions. This represents a false positive rate of 7%. The two false alerts were caused by the incompleteness of our static database. Everyone was a special case that we did not anticipate. The last false positive was caused by an application that integrates a website using HTML5 geolocation, which requires a location permission. We wrote test cases for these scenarios and updated our permissions database.

Code Analysis Results: Dynamic analysis shows that 19% of all PlayStore apps contain dynamic class loading (DCL) calls. 8% of BUSINESS applications, 10% of SHOPPING apps and 10% of TRAVEL AND LOCAL apps have the minimum percentage, however, 39% of GAME apps are using dynamic class loading feature. The results can be easily understood. in recent years, games for mobile platforms have evolved considerably. This forces a developer to write tons of code for different versions of the mobile platform. It is therefore not surprising that developers choose a strategy when an original application (whose size is limited to 50 MB) is only an installer, which downloads additional code from a server and loads it dynamically. On average, there are 108 dynamic class load function calls per application. Considering reflection, 78% of all applications use reflection calls that are interesting for our dynamic analysis. The percentage reaches 98% for applications that belong to the GAME category, which means that almost every application in this category is based on reflection. In addition, applications in this category also show the highest average number of reflection calls (38.13), 21.96 reflection calls per application.

4 Conclusion

In this paper, we present a Tool called "PerUpSecure" that analyzes permissions requested by Android applications in installing-time and after their updates. Our reference implementation is very efficient and induces a small performance overhead. Therefore, we have developed this tool especially for users without a technical and security background. We analyzed a large set of normal and malicious applications collected from different applications market. The majority of this article provide the results we got.

References

1. Deloitte: State of the smart Consumer and business usage patterns. Mobile, Glob. Surv. Consum. Cut, UK (2017)
2. IDC: Smartphone market shares (2017). https://www.idc.com/promo/smartphone-market-share/os
3. Suarez-tangil, G., Stringhini, G.: Eight Years of Rider Measurement in the Android Malware Ecosystem: Evolution and Lessons Learned, arXiv, pp. 1–18 (2017)
4. Rajesh, B., Reddy, P., Patil, M., Pareek, H.: Droidswan: detecting malicious android applications based on static. In: Conference *on* Communications Security & Information Assurance, May 2015
5. Ahvanooey, M.T., Li, P.Q., Rabbani, M., Rajput, A.R.: A survey on smartphones security: software vulnerabilities, malware, and attacks. Int. J. Adv. Comput. Sci. Appl. **8**(10), 30–45 (2017)
6. Sarma, B., Li, N., Gates, C., Potharaju, R., Nita-rotaru, C., Lafayette, W.: Android Permissions: a perspective combining risks and benefits, pp. 13–22 (2012)
7. Reshetova, E., Karhunen, J., Nyman, T., Asokan, N.: Security of OS-level virtualization technologies. In: Conference Security, IT (2014)
8. Skillen, A., Van Oorschot, P.C.: Deadbolt: locking down android disk encryption*. In: Proceedings of the Third ACM Workshop on Security and Privacy on Smartphones Mobile Devices, pp. 3–14. ACM (2013)
9. Wijesekera, P., Columbia, B., Baokar, A., Hosseini, A., Egelman, S., Wagner, D.: Android permissions remystified: a field study on contextual integrity T. In: USENIX Security Symposium, pp. 499–514 (2015)
10. Chen, K.Z., Johnson, N., Silva, S.D., Macnamara, K., Magrino, T., Wu, E., Rinard, M., Song, D.: Contextual policy enforcement in android applications with permission event graphs. In: NDSS (2013)

11. Felt, A.P., Chin, E., Hanna, S., Song, D., Wagner, D.: Android permissions demystified. In: Proceedings of 18th Conference on Computer & Communications Security - CCS '11, p. 627 (2011)
12. Zhang, X., Breitinger, F., Baggili, I.: Rapid Android Parser for Investigating DEX Files (RAPID), vol. 17, pp. 28–39 (2016)
13. Barros, P., Millstein, S., Vines, P., Dietl, W., Amorim, M., Ernst, M.D.: Static analysis of implicit control flow: resolving Java reflection and android intents. In: 2015 30th IEEE/ACM International Conference on Automated Software Engineering (ASE) (2015)
14. Ding, J.: PQEMU : A parallel system emulator based on QEMU. In: 2011 IEEE 17th International Conference Parallel Distributed Systems (2011)
15. A. Developers: UI/Application Exerciser Monkey. August, 2012. http://developer.android.com/%0Atools/help/monkey.html
16. G. Play: Android official market. https://play.google.com/store/apps
17. A. Market: AndroidBest. http://androidbest.ru/
18. A. Market: AndroidDrawer. http://www.androiddrawer.com/
19. A. Market: AndroidLife. http://androidlife.ru/
20. A. Market: Anruan. http://www.anruan.com/
21. A. Market: AppsApk. http://www.appsapk.com/
22. A. Market: PandaApp. http://android.pandaapp.com/
23. A. Market: SlideME
24. Barrera, D., Van Oorschot, P.C., Somayaji, A.: A methodology for empirical analysis of permission-based security models and its application to android. Security **1**, 73–84 (2010)
25. Enck, W., Ongtang, M., Mcdaniel, P.: Understanding android security. IEEE Secur. Priv. (1), 50–57 (2009)

Effect of Climate Change on Growth, Development and Pathogenicity of Phytopathogenic Telluric Fungi

Mohammed Ezziyyani[1(✉)], Ahlem Hamdache[2], Meryem Asraoui[3], Maria Emilia Requena[4], Catalina Egea-Gilabert[5], and Maria Emilia Candela Castillo[4]

[1] Polydisciplinary Faculty of Larache, Department of Life Sciences, Abdelmalek Essaâdi University, 745 Poste Principale, 92004 Larache, Morocco
mohammed.ezziyyani@gmail.com

[2] Faculty of Sciences of Tetouan, Department of Biology, Abdelmalek Essaâdi University, Avenue de Sebta, Mhannech II, 93002 Tétouan, Morocco

[3] Faculty of Sciences and Techniques of Settat, Hassan Premier University, Route de Casablanca Km 3, 5 BP 539 Settat, Morocco

[4] Faculty of Biology, Department of Plant Biology, University of Murcia, Campus de Espinardo, 30100 Murcia Murcia, Spain

[5] Department of Science and Agrarian Technology, Agronomic Engineering, University Politenic of Cartagena, Paseo Alfonso XIII, 48, 30203 Cartagena, Spain

Abstract. Soil microorganisms are extremely numerous and diverse. This diversity responds to the multitude of biogeochemical microenvironments of the soil as well as to the complexity of the forms of organic matter in the soil, their energy resource. Their distribution in the soil is very heterogeneous and is explained by the presence of conditions supporting the development of life. A very likely consequence of global warming would be a change in the range of some phytopathogens such as *Phytophthora capsici*, *Rhizoctonia solani* and *Fusarium oxysporum*. The fungi live in relatively homogeneous conditions. They are all heterotrophic microorganisms living under aerobic conditions. Indeed, certain microorganisms are known to have a distribution limited by temperature. To do this, we focused on the mean rate of mycelial growth as a function of the time (V_{max} = d/t) of the three phytopathogens, at three different temperatures (20, 25 and 30 °C) and we also used a series of agroclimatic indices. The results show that *F. oxysporum* and *R. solani* have a very limited distribution at 22 and 30 °C ($V_{max} \approx 10$ mm) for 72 h; however *P. capsici* showed a $V_{max} \approx 20$ mm for 72 h, although the pathogen also depends on the temperature, probably its reproductive success as well as its distribution and speeds of development are extremely related to moisture. The pathogenicity analyzed by artificial inoculation of pepper seedlings shows that *P. capsici* is very aggressive at 30 °C, *F. oxysporum* showed virulence only at 25 °C but *R. solani* lost all virulence between 22 and 30 °C.

Keywords: Climate change · Temperature · Telluric phytopathogenic fungi

M. Ezziyyani (Ed.): AI2SD 2018, AISC 911, pp. 14–21, 2019.
https://doi.org/10.1007/978-3-030-11878-5_2

1 Introduction

Agriculture occupies an important place in the Moroccan economy and is largely influenced by climatic conditions. This is all the more true since several Moroccan regions are at the northern limit of agricultural production. In addition, climate change will be greater in northern Morocco than in the south. These projected climate changes are likely to have negative impacts on agriculture and sustainable development. The impacts of climate change on agriculture and human well-being include biological effects on crop yields, resulting downstream impacts, including those on prices, production and consumption, and impacts on calorie consumption per capita and on child malnutrition. The year 2015 was marked by meteorological events such as floods, drought and heat making this year the hottest in history according to NASA (National Aeronautics and Space Administration). NASA [1] has released a video that reflects the unstoppable increase in global average temperature from 1880 to 2015. The images show an indisputable trend of global warming. The problem lies not only in our perception of heat, but in the increase of natural disasters such as drought, heat waves, floods, electrical storms. Without mentioning the effects that this causes in the melting of the poles, in the fauna and the flora. Although pathogens also depend on temperature, their reproductive success and rate of development are extremely moisture-related. The agricultural sector will have to adapt to climate change in a variety of ways. One of the constraints it will face is crop pests, since the biology of pests is very sensitive to climate change. In this context, anticipated climate change by 2050 will potentially affect crop/pest relationships [2]. It is essential to realize right now that the pressure caused by pests will change as the climate changes. This is particularly true in the context where the industry is seeking to reduce the use of pest control products that have adverse effects on the environment and human health. Temperature and humidity are the main bioclimatic factors influencing the development of diseases, while CO_2 and O_3 have an indirect effect on pathogens via the physiology of the plant [3]. The softening of winters will, in general, ensure a better preservation of pathogens, thus increasing the amount of soil inoculum the following spring [4]. The biophysical effects of climate change on agriculture lead to changes in production and prices, which in turn affect the economic system as farmers and other market participants adapt individually changing crop choice, input use, production, food demand, food consumption and trade. It is increasingly evident that climate change is happening. The international scientific community is unanimous about climate change: it is a real phenomenon whose effects are already being felt in certain regions. In addition, it is widely recognized that climate change will intensify despite the implementation of important measures to reduce greenhouse gas emissions and will have economic, social and environmental consequences for climate change for Morocco and its inhabitants. Although the effects vary from one region to another, all regions and virtually all economic sectors in the country will be affected. It is especially the changes in climate, temperature or rainfall that are likely to have direct consequences on pests and insect pests. In return, the increase in the CO_2 content of the air does not seem to have a direct influence on the development of parasitic fungi and bacteria [5] or on that of insects [6]. The epidemic development of many parasites is strongly influenced by the conditions of temperature and humidity, whether for dissemination, infection or multiplication. This is particularly the case of many leaf parasites that are highly dependent on the moistening of leaves for infection or splashing caused by rain for dispersion (rust or *Marssonina* on poplars). Others, such as oak powdery mildew, are favored by low humidity. The epidemics provoked by these parasites have a very variable importance according to the years. It is clear that long-term changes in rainfall or higher temperatures would have an impact on the importance of epidemics, although it is very difficult to predict which diseases would be favored or disadvantaged. However, we can expect more or less significant modifications of the parasite processions that we know. The rise in temperature also affects the processions of parasites and predators of insect pests [7]. Since their thermal optima may be different, global warming may be favorable to insect pests [8], and sometimes favorable to their natural enemies [9]. A development of the insect pest accelerated compared to that of its parasite or predator could also allow it in certain cases to escape from its aggressor, and thus to benefit from a temporal refuge effect [10].

2 Materials and Methods

2.1 Pathogenic Fungi

The selected virulent pathogens are *Phytophthora capsici*, *Rhizoctonia solani* and *Fusarium oxysporum*, obtained from the rhizosphere of pepper plants showing symptoms of wilting and crown rot. Their pathogenicity has been experimentally confirmed in vivo on leaves, stems and fruits of peppers. To avoid the development of secondary fungi and bacteria, the *pythiaceae* selective medium developed by Ponchet [11] was used for all isolates of the genus *Phytophthora*; on two different backgrounds: Acidified PDA (pH = 4) with citric acid by Komada [12] selective for *Fusarium* genus and selective agar medium ($K_2H\ PO_4$: 1 g, Mg SO_4, 0.5 g, KCl: 0.5 g, Fe SO_4: 0.01 g, Na NO_2: 0.2 g, tannic acid: 0.12 g, propamocarb (Schering): 0.8 g, streptomycin sulfate: 0.1 g, chlortetracyclin: 0.05 g, etridiazole (Seppic): 0.005 g, $CuSO_4$: 0.0013 g, benomyl (Dupont): 0.0013 g, agar 15 g, H_20 qsp: 1000 ml, sterilization 30 min/110 °C) for put the *Rhizoctonia* genus in prominence Camporota [13]. The leaves, roots and neck of the sample are washed, dried and then deposited on each selective medium. After confirmation of their membership, the strains, stored in tubes containing PDA medium at room temperature of the laboratory, are sub-cultured every 3 or 4 weeks.

2.2 Temperature Tests

The rate of mycelial growth of pathogens was studied at three different temperatures (20, 25 and 30 °C) on a PDA culture medium (Difco, Detroit, Michigan, USA). The experiments were conducted in 90 mm Petri dishes inoculated with a 3 mm diameter mycelial implant. Implants containing mycelium from each isolate were placed in direct contact with the culture medium. Three boxes were used for each isolate and temperature. The experiment was repeated three times simultaneously. The boxes containing the mycelial implants were incubated in the dark for 7 days. Two perpendicular diameters per colony were measured daily without opening the Petri dishes and the growth rate was calculated on the basis of these two diameters.

2.3 Statistical Analysis

The diametral growth and sporulation values of different phytopathogens *P. capsici, R. solani* and *F. oxysporum* on culture medium are compared statistically by LSD (Least significant Difference), multiple range test ($P < 0.05$).

2.4 Results and Discussion

The results shown in Table 1 show that the averages of diametral growth and sporulation intensity (Fig. 1) of *P. capsici*, *R. solani* and *F. oxysporum* on PDA medium show significant differences at the fold level 5%. After three days after incubation for 3 days at 20, 25 and 30 °C, on agar medium. The length of *P. capsici* mycelium was 2 cm at 20 °C, 2.9 at 25 °C and 4.2 °C at 30 °C. *F. oxysporum* was 1 cm at 20 °C, 2.2 cm at 25 °C and 0.8 cm at 30 °C. *R. solani* was 1.3 cm at 20 °C, 1.5 cm at 25 °C and 1 cm at 30 °C. Mycelial growth and spore production is good at 30 °C for *P. capsici*, medium between 25 and 30 °C for *F. oxysporum* and almost nil between 25 and 30 °C for *R. solani*.

Table 1. Diametral (cm) growth of mycelial phytopathogens on agar medium

Phytopathogens	20 °C	25 °C	30 °C
P. capsici	2.0	2.9	4.2
F. oxysporum	1.0	2.2	0.8
R. solani	1.3	1.5	1.0

On a global scale, average surface temperatures increased by about 0.6 °C during the 20th century. According to the climate scenarios summarized by the Intergovernmental Panel on Climate Change (IPCC), average global temperatures are projected to rise from 1.4 to 5.8 degrees C in this century. It should be noted that the four hottest years (2017 being limited to January-August) are in the order 2016, 2017, 2015 and 2014. 2017 will therefore be the hottest year if the average over the last four months is higher than at +1.14 °C. It's pretty unlikely. 2017 will be the second hottest year if the average over the last four months is greater than +0.75 °C. This could happen if the current trend continues. 2017 will be the hottest third year if the average over the last four months is lower +0.75 °C. This is not excluded, knowing that September, October, November and December the 0.75 °C have been exceeded only three times in instrumental history. However, there is no chance to see 2017 fall to fourth place: it should be an average of +0.33 °C, which has not happened since 2000.

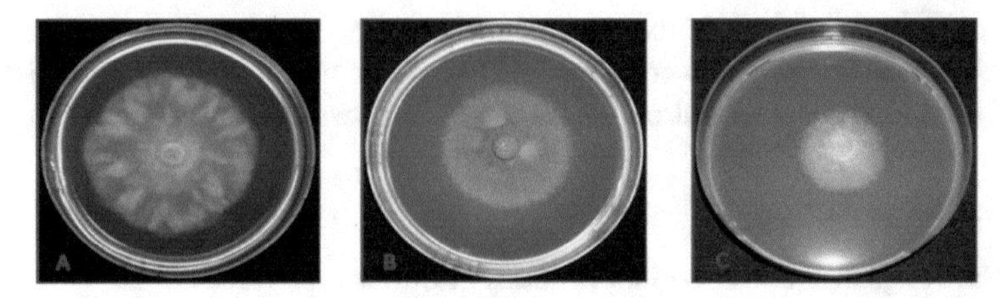

Fig. 1. *In vitro* growth at 25 °C of *P. capsici* (A), *F. oxysporum* (B) and *R. solani* (C), in agar medium

The impact of climate change is more difficult to predict with these organisms since the uncertainty related to rainfall forecasts in the near future is greater than for temperature averages. Nevertheless, several studies show that climate change will have a positive effect on some species by increasing their establishment capacity, their growth rate and the duration of the epidemic [3]. Climate extremes also have a role to play in the infectivity of a pathogen by promoting its entry, for example, by a wound to the plant caused by hail, drought or frost. The resistance of the cultivated plant can thus be compromised by these wounds or by physiological or morphological changes, as mentioned above [14]. For example, the expected future decline in snow cover in southern Quebec and its isolation capacity may, however, be detrimental to certain cereal diseases, for example, which could see their winter survival rate drop. Soil-dwelling genera (*Botrytis*, *Fusarium*, *Phytophthora*, *Pythium*, *Rhizoctonia*, *Sclerotinia*, *Sclerotium*, *Verticillium*) are less likely to be affected by this phenomenon due to survival structures such as sclerotia or hyphae [3]. Warm periods with droughts may reduce the intensity of some pathogens, but may also reduce plant resistance to disease [15]. Some species such as *Podosphaera*, *Sphaerotheca*, *Uncinula* and *Ustilago* could however benefit from a drier and hotter climate. A study of vine blight, a disease developing at cooler temperatures (10 to 24 °C), predicts an increase in the intensity of infections with an increase in temperature and a decrease in precipitation [16, 17]. An expansion of the range of pathogens is also to be expected [18] in connection with the increase in temperatures. The extension of the growing season duration would allow greater inoculum production for some species and increase the frequency and intensity of infections. Since moisture is a major factor in the development of most fungal infections, increased precipitation favored spore dispersal of some species. The amount of precipitation accumulated in southern Quebec, however, should not vary significantly by 2050 for the summer period [19]. Further work on the influence of temperature on the different life stages of *Curvularia lunata*: Mycelial growth and sporulation of isolates are average at 20 °C and optimal at 28 °C. According to several authors [20, 21], temperatures between 15 and 33 °C are optimal for the growth of the *Curvularia* sp mycelium. However, some species of this genus can develop over a wide range of temperatures [22]. At 35 °C, mycelial growth and sporulation of *C. lunata*, although small, are still possible. But after exposure of cultures to temperatures of 5 and 40 °C, both growth and sporulation are inhibited. The inability of certain fungi to grow at high temperatures [23] may be directly related to their inability to synthesize, at these temperatures, substances necessary for growth, such as vitamins [24].

3 Conclusion

P. capsici, *R. solani* and *F. oxysporum* have a short survival time at temperatures below 0 °C and moreover it is mostly active at temperatures of 22 to 25 °C for *R. solani*, from 24 to 27 °C. C for *F. oxysporum* and 30–35 °C for *P. capsici*. Its thermal requirements probably limit their current spatio-temporal distribution and their virulence. Today, most of the research on impacts and adaptation has focused on the biophysical effects of climate change. The results suggest that the most important challenges will be related to the increasing frequency and intensity of extreme weather events such as floods, droughts and storms. In contrast, many systems should be able, subject to appropriate adjustments, to tolerate a gradual and limited warming of temperatures, and even to take advantage of them at times. Thus, in some areas, the increase in temperature could promote and/or inhibit the growth of pathogens. Although plant pathogens encompass several types of organisms, this section focuses mainly on fungal infections, dominant in Morocco in our case in the Larache Region. There are also gaps in research on the impact of climate change on bacteria, viruses and nematodes, enemies that can cause considerable damage to crops while the methods of struggle are almost non-existent. Climate change will have an effect on plant physiology, which may in turn alter the passive resistance of plants to disease. Physiological or physical changes in the plant, related to climate change, may decrease or increase susceptibility to pathogens. An increase in the incidence of diseases and plant susceptibility could potentially limit the range of crops/cultivars available in the future. As far as fungicides are concerned, their effectiveness in controlling pathogens is likely to be affected by temperature increases that could reduce their toxicity or frequent rainfall events that increase leaching before they enter the plant. We report a few studies have focused on modeling the impacts of climate change on pathogens. In fact, they need moisture at different stages of their life cycle and the modeling of precipitation and especially of relative humidity presents a greater range of uncertainty than temperature modeling. This uncertainty is reflected in a range of potentially variable biological responses. In addition, the biology of pathogens is also very different from one species to another in terms of their responses to climatic parameters, making it difficult to establish generalities about the impact of climate change on climate change on these. The agricultural sector will have to adapt to climate change in a variety of ways. One of the constraints it will face is crop pests, since the biology of pests is very sensitive to climate change. In this context, anticipated climate change by 2050 will potentially affect the crop / pathogen relationships of crops. It is essential to realize right now that the pressure caused by crop pathogens will change with climate change. This is particularly true in the context where the sector is seeking to reduce the use of pathogen control products that have adverse effects on the environment and human health.

References

1. NASA: National Aeronautics and Space Administration (2017)
2. Trumble, J.T., Butler, C.D.: Climate change will exacerbate California's insect pest problems. Calif. Agric. **63**, 73–78 (2009)
3. Boland, G.J., Melzer, M.S., Hopkin, A., Hihhins, V., Nassuth, A.: Climate change and plant disease in Ontario. Can. J. Plant Pathol. **26**, 335–350 (2014)
4. Fuhrer, J.: Agroecosystem responses to combinations of elevated CO_2, ozone, and global climate change. Agric. Ecosyst. Environ. **97**, 1–20 (2003)
5. Manning, W.J., Tiedeman, A.V.: Climate change: potential effects of increased atmospheric carbon dioxide (CO_2), ozone (O_3) and ultraviolet-B (UV-B) radiation on plant diseases. Environ. Pollut. **88**, 219–245 (1995)
6. Fajer, E.D., Bowers, M.D., Bazzaz, F.A.: The effects of enriched CO_2 atmospheres on the Buckeye Butterfly. Junonia Coenia. Ecology **72**, 751–754 (1991)
7. Mattson, W.J., Haack, R.A.: The role of drought in outbreaks of plant-eating insects. Bio Sci. **37**(2), 110–118 (1987)
8. Fleming, R.A., Volney, W.J.A.: Effects of climate change on insect defoliator population processes in Canada's boreal forest: some plausible scenarios. Water Air Soil Pollut. **82**, 445–454 (1995)
9. Skirvin, D.J., Perry, J.N., Harrington, R.: The effect of climate change on an aphid-coccinellid interaction. Glob. Change Biol. **3**, 1–11 (1997)
10. Hassell, M.P., Godfray, H.C.J., Comins, H.N.: Effects of global change on the dynamics of insect hostparasitoid interactions. In: Biotic Interactions and Global Change, pp. 402–423. Sinauer Associates Inc. (1993)
11. Ponchet, J., Ricci, P., Andreolli, C., Auge, G.: Méthodes sélectives d'isolement du Phytophthora nicotianae fsp parasitica (Dastur) Waterh à partir du sol. Ann Phytopathol. **42**, 97–108 (1972)
12. Komada, H.: Development of a selective medium for quantitative isolation of Fusarium oxysporum from natural soil. Rev. Plant Prot. Res. **8**, 114–124 (1975)
13. Camporota, P.: Mesure de la colonisation saprophytique en compétition de Rhizoctonia solani Kühn dans les sols et substrats. Agronomie **1**(6), 531–5177 (1981)
14. Rillig, M.C.: Climate change effects on fungi in agroecosystems. In: Newton, P.C.D., Carran, R.A., Edwards, G.R., Niklaus, P.A. (eds.) Agroecosystems in a Changing Climate (2015)
15. Gregory, P.J., Johnson, S.N., Newton, A.C., Ingram, J.S.I.: Intergrating pest and pathogens into the climate change/food security debate. J. Exp. Bot. **60**, 2827–2838 (2009)
16. Salinari, F., Giosuè, S., Tubiello, F.N., Rettori, A., Rossi, V., Spanna, F., Rosenzweig, C., Gullino, M.L.: Downy mildew (Plasmopara viticola) epidemics on grapevine under climate change. Glob. Change Biol. **12**, 1299–1307 (2006)
17. Del Ponte, E.M., Fernandes, J.M.C., Pavan, W.M., Baethgen, W.E.: A model-based assessment of the impacts of climate variability on Fusarium head blight seasonal risk in southern Brazil. J. Phytopathol. **157**, 675–681 (2016)
18. Coakley, S.M., Scherm, H., Chakraborty, S.: Climate change and plant disease management. Ann. Rev. Phytopathol. **37**, 399–426 (1999)
19. Bao, J.R., Zhan, Y.C., Wu, X.S., Yu, S.F.: Studies on the pathogen of stem rot of chinese waterchestnut and its biology. Acta Physiol. Sin. **20**, 311–370 (1993)
20. Curtis, R.: Some host parasite relationships in the Curvularia disease of Gladiolus in Florida. Plant Dis. Rep. **45**, 512–516 (1961)

21. Desjarlais, C., Allard, M., Belanger, D., Blondlot, A., Bouffard, A., Bourque, A., Chaumont, D., Gosselin, P., Houle, D., Larrivee, C., Lease, N., Savard, J.-P., Turcotte, R., Villeneuve, C., Montreal, QC.: Savoir s'adapter aux changements climatiques. Ouranos (2010)
22. Dhawan, S., Pathak, N., Gary, K.L., Mishra, A., Agrawal, O.P.: Effect of temperature on some fungal isolates of Ajanta wall paintings. In: Proceeding of the International Conference on Biodeterioration of Cultural Property, Lucknow, India, pp. 339–352 (1991)
23. Mishra, R.R., Pandey, K.K.: Studies of soil fungistasis: V. Effect of temperature, moisture content and inoculation period. Indian Phytopathol. **27**, 475–479 (1974)
24. Moore-Landecker, E.: Fundamentals of the Fungi. 2nd edn., 578 p. Prentice-Hall, Englewood Cliffs (1982)

An Intelligent Approach for Enhancing the Agricultural Production in Arid Areas Using IoT Technology

Abdelhak Merizig(✉), Hamza Saouli, Meftah Zouai, and Okba Kazar

LINFI Laboratory, Computer Science Department,
University of Biskra, Biskra, Algeria
merizig.abdelhak@gmail.com, hamza_saouli@yahoo.fr,
zouaimeftah@yahoo.fr, kazarobka@gmail.com

Abstract. The increasing of dates fruit development in Algeria becomes important for the next generations because it can enhance the national economy. To improve the production of this treasure more than more, we need to analyze and to monitor the previous production for giving the consequences that can make the best results. Since we have many farms with a large number of palms tree it is a difficult task to supervise and collect data in a short time. For this major problem, we need to integrate a set of components that can communicate together to support the farmers in collecting data. Therefore, the solution to this issue is to propose an intelligent architecture that uses a method that can help the expert to make decisions. In this work, we present a solution to forecast the dates fruit production based on historical data, in order to enhance the quality and the performance of the production in coming years. Moreover, to collect data in this work we use an intelligent technology with a drone to facilitate the collection operation. The implementation of this model has been provided in order to evaluate our system. The obtained results demonstrate the effectiveness of our proposed system.

Keywords: Agricultural science · Dates fruit production · IoT agriculture · NFC · Support vector regression · Decision support system

1 Introduction

Nowadays, the production becomes an important factor for the productive country, because it gives the importance to a country to be one of the sophisticated countries. For long time providers and farmers attempt to find a solution that able to make decisions in order to help them to improve the quality of production. In addition, in order to enhance the production, we need a system that can support the experts to make decisions in order to provide precautions to the farmers to use it in exceptional weather cases. To improve this production, we need an intelligent system to automate research tasks and respond in a reasonable manner using machines. Moreover, the farms are known by a large space it might be a hundred of hectares. These large surfaces make data collection operation more difficult especially in pays that have a difficult weather condition either the hotter one or the colder one. Therefore, Informatics Technologies

M. Ezziyyani (Ed.): AI2SD 2018, AISC 911, pp. 22–36, 2019.
https://doi.org/10.1007/978-3-030-11878-5_3

(IT) comes with their methods and algorithms to solve problems in the real life. The famous and traditional one of this approach is a decision support system (DSS). The decision support system came to help users to make decisions in specifying the problem, in the process of decision-making [1], this tool offers suppliers results that can improve their production. The aim of this work it consists to give a system which can offer to users monitoring the production of dates fruits illustrated by graphical schemes. Furthermore, it gives forecasts of production for the coming years using two methods that are support vector machines for regression (SVR) and exponential smoothing based on historical production. Furthermore, our solution proposed in this paper is to install sensors in some palms tree using the k-means method. Using drone can help the farm owner to collect data in a short time. To achieve the principal goal that makes the production better than previously.

This paper is organized as follows: Sect. 2 outlines some related work. Section 3 we give an example of the case study. Section 4 presents some related concepts used in our work. In Sect. 5, we present our approach and provide the description of the proposed architecture. In addition, Sect. 5 presents the used algorithm to construct the prediction model. Section 6 shows the implementation and the obtained results using our prediction models. Finally, Sect. 7 concludes this paper.

2 Related Works

Production's forecasting represents a challenge for the economic experts due to the importance of that will give this operation. In the agricultural field, it a necessary to find a good model to predict the production to avoid a big lost compared to previous years. In this section, we are going to give some works that tried to solve the problem of production forecasting.

Bhardwaj et al. [2] proposed a solution for wheat production using a model based on structural time series. In addition, the proposed model used a real data collected during five years. The authors have proved some results using "Structural time-series modelling (STSM)" which to avoid ARIMA method limits. However, the factors that can affect these forecasts are not considered as the climatic factors.

Deshmukh and Paramasivam [3] presented in their work a solution based forecasting method to predict the milk production in India. The presented problem can enhance the growth and development sector. The authors have used a real value collected form twenty to fifty years. Moreover, Deshmukh and Paramasivam use statistical method untitled ARIMA to construct a model in order to forecast the production. However, like the previous work including some factor that affects the production can improve the prediction more that more.

Amin et al. [4] proposed a solution that could enhance the production of wheat, which represent a big challenge in Pakistan. In addition, this operation could determine the production situation for the coming year. The authors developed a time series model based on ARIMA method. However, including some other factors could enhance the prediction task.

3 Case Study

As known to everyone in the agricultural field, the trees are located everywhere in a large space area that might be thousands of hectares. Through this important space makes the data collecting difficult due to the weather conditions (that may be heating or windy) or a long distance to walk in order to collect data. These data collected by the farmer consists of climatic factors or the number of fruit in each tree. Furthermore, Algeria is known for the quantity and the quality of dates fruit, in this paper we are going to introduce a system that can provide forecasts for the production of dates. Moreover, forecast operation it is necessary and can help any farmer's owner to prepare some precautions to avoid the damage of fruits during the year and consequently, it increases the production of this treasure.

4 Background

In this section, we present some definition and technologies used in this paper since we are in the agricultural field.

1 **NFC technology.** Near Field Communication (NFC) is a technology that facilitates communication between two different devices (NFC-enabled devices) by using low power and radiating signals around a close proximity about 4 cm [5].
2 **Drones** are a sort of robots that can autonomously fly in the different area sometimes have sensors and cameras. These drones are used in many applications like military, agricultural and so on [6].
3 **Sensors** are devices that able to detect, receive any information or signals from the environment, and they are composed of a set of electronical components such as a camera. We have many types of sensors like light, gas, motion, pressure and so on [7, 8].

5 Proposed Approach

To provide a system that can give the prediction of production this section present the proposed architecture to handle the problem at hand. In addition, the proposed architecture is based on a decision support system (DSS) architecture type [9]. Moreover, as we mentioned earlier this work offer results to both users either the simple client or an expert through displaying results (using graphical forms or giving the production forecasts for the next years). Next Fig. 1 shows the proposed architecture.

5.1 Architecture Description

In this subsection, we are going to present the general description of our proposed architecture. Moreover, we will explain the role of each component. The proposed architecture aims to present a solution to the farmers and the experts in a way that helps

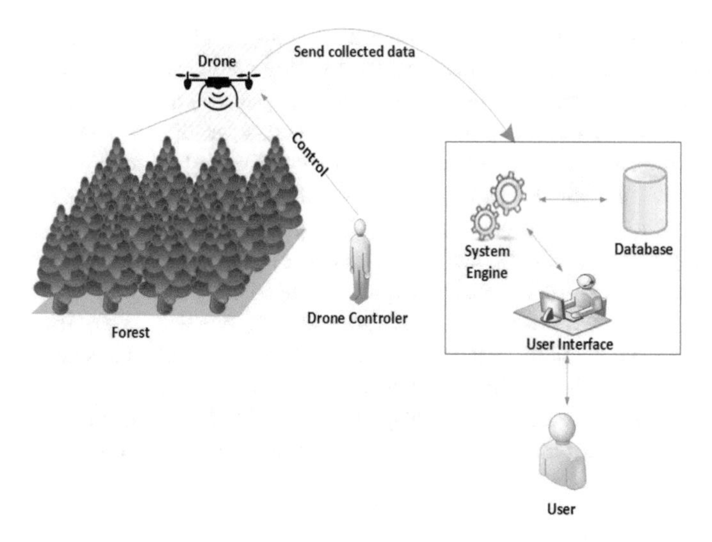

Fig. 1. The proposed architecture for production forecasts.

them to enhance the production of dates. As Fig. 1 depicts, the proposed architecture is composed of the following components:

1. **User interface:** this is the component represents the link between the internal and the external world (clients and the system) using a graphical interface. Furthermore, the purpose of this component is to collect the information from the user and to display results. This interface provides three functions which are:

   – **Simple research.** The clients of any field generally interest this type of research, for the culture on dates of different regions exists.
   – **Search with graphic shapes.** This function is destined for experts in the production of dates field either a decision maker or analyzer. According to their needs, the interface displays these results as curves, stacked column, pie chart and tables.
   – **Forecasting.** This operation is given to decision makers; their aim is to make a study in coming years based on existing data, in order to give an advantage to the businesses through obtained results.

In case the user is an administrator, he can apply some changes to the available data using the next operation.

– **Filling data.** The role of this operation is to fill the database from the administrator interface since statistics already made on dates (quantity and variety).
– **Update data.** In this function, the administrator can modify existing data on the database. In addition, he can also remove one or more information from the database.

2. **Database**: this component plays the role of storage data in our system, this data is stored by the administrator according to existing statistics. In addition, each record contains the data about the production, the total palmers number and the productive palmers number, and each information has data: year, city, wilaya and cultivar.
3. **System engine**: this component is the kernel of the system; it is the responsible of all available operations. For every question chosen by the user either simple or with graphic schemes or production forecasting, the system answer for questions according to the available data in the database.
4. **Drones and sensors:** these components are installed in our architecture to add the intelligent part in the field of agricultural. In fact, the sensors are located in some places between the palms tree in order to collect the data. The considered data in our work are linked to weather factors which include: temperator, humidity, wind speed, and precipitation. As displayed in the Fig. 1 we assign a person (drone controller) who can manipulate the drone in order to get the information using NFC technology.

5.2 System Functionality

The presented proposition gives to the users (clients or expert) a system that can give some forecasts for the next years in order to enhance the national production. In addition, it provides some statistical data to study some scenarios. The next flowchart shows the different steps used in the proposed system see Fig. 2.

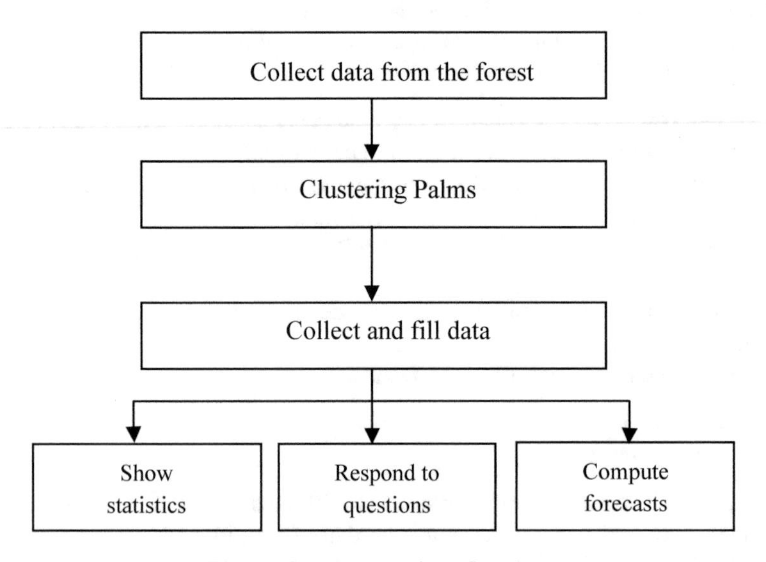

Fig. 2. System operations flowchart.

- **Collect data from the forest.** In this step, the drone collects the information about the number of palms tree in order to calculate the number of sensors to be installed for each part of the forest.

– **Clustering palms.** After calculating the number of palms tree in the forest, this step we are going to define the number of the cluster which includes a set of palms tree to install the sensors in order to collect the data about the weather. These climatic factors include the temperature, wind speed, humidity and precipitation. In the literature, there are many methods to divide a set of data, in this work we have used K-means method [10] to construct a set of the cluster which is in our case as set for palms tree in order to install the sensor in the center of each cluster. We have used this method in order to reduce the number of sensors because they are expensive to put it on each palms tree. In fact, a k-means method is based on partitioning the data into a sort of data in particular form [11]. At the end of this algorithm, we aim to minimize the distance between the center of each cluster. The configuration of the algorithm defined [9] and presented as follows:
  - n number of sensor nodes $S_i : i \in [1, 2..., n]$ in k partitions. $P_j \in [1, 2..., k] (k \leq n)$ from k centers chosen arbitrary "C_j".

$$\arg min_p \sum_{j=1}^{k} \sum_{i=1}^{n} \|S_i - C_j\|^2 \tag{1}$$

Where $\|S_i - C_j\|^2$ is the distance between a sensor 'S_i' and the partition center 'C_j'.

After calculating and achieving the best the distance in each cluster, in the center of these clusters we install the sensors in each center. Figure 3 shows the installed sensors in each cluster.

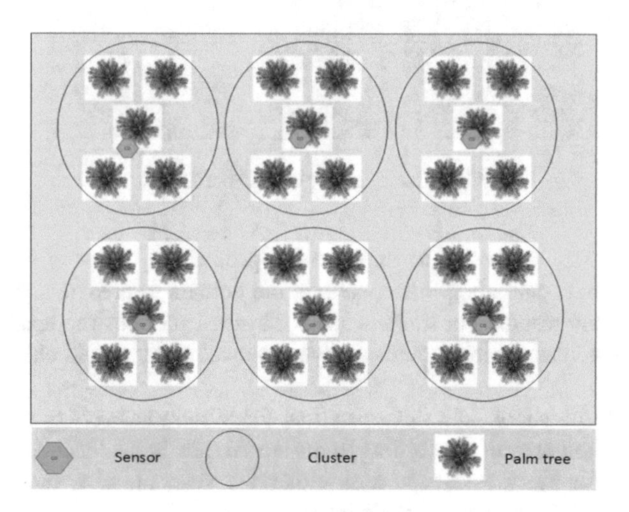

Fig. 3. Clustering using the k-means method.

– **Collect and fill data.** In this step, the drone controller uses the information collected by the sensors using the NFC technology which allow it to communicate with the drone. When we get the data, we send it directly to the system in order to store them in the local database. After this step, we can use the existed data in order to serve the clients. We have created three tables in our database which contains the information about the date fruits per year. According to the Directorate of Agricultural Services at Wilaya of Biskra, these data are divided into three parts:

- A total number of palms tree.
- A number of productive palms tree.
- The date fruits production.

In order to collect all the data from the field, the drone controller follows a scheme described in the next Fig. 4.

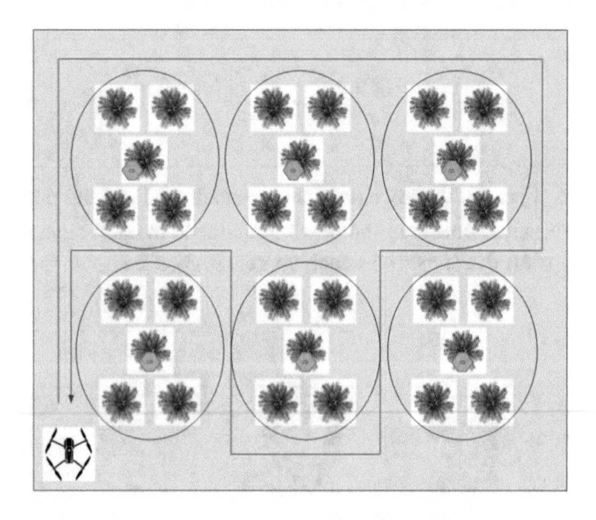

Fig. 4. Mapping plan for the drone.

- **Show statistics** – In order to improve the production in the business economy, statistical results could help the experts in the domain to prepare a study according to the obtained results. For this matter, our system presents the results in different graphical schemes, which includes curves, stacked column, pie chart, and a set of tables.
- **Respond to questions** – In this operation, the system answers to the needs of the user as question-answer. According to the stored data in the local database, the user enters data for every asked question either the year, the city, the wilaya, and/or cultivar from a form exists at their interface.

- **Compute forecasts** – This operation allows for giving a prediction on the dates production during the coming years basing on previous years, in our work we have used two prediction methods presented as follows:

1. **The Holt's exponential smoothing method**

The method of Holt's exponential smoothing, also called the Holt-Winters method with no seasonal trend, use the prediction function [12]:

$$\hat{X}_{T+h} = b_T + ha_T \tag{2}$$

To prediction $\hat{X}_{T+h}$ from the observations $X_1, X_2, \ldots, X_T$.

$\hat{X}_{T+h}$: represents the prediction.
a_T: represents smoothing trend (also called the slope).
b_T: represents smoothing average (also called the level).
h: represents the forecast horizon.

The level a_T and the slope b_T are updates applying the following formulas:

$$\begin{cases} b_T = \alpha X_T + (1-\alpha)[b_{T-1} + a_{T-1}], 0 < \alpha < 1, \\ a_T = \beta[b_T - b_{T-1}] + (1-\beta)a_{T-1}, 0 < \beta < 1. \end{cases} \tag{3}$$

A reasonable choice of initial estimates is to set the initial values in a_0 and b_0:

$$b_0 = X_1, \text{ and } a_0 = 0 \tag{4}$$

To choose the two constants α and β , we must use the minimization of errors method to specify the coefficients α and β:

$$\text{Error} = \sum_{t=1}^{T-N} \left[X_t - \hat{X}_t\right]^2 \tag{5}$$

T: the total number of instances.
N: the number of instances of the test.

The next Algorithm 1 shows the different used steps to implement the Holt's exponential algorithm which helps us to construct the predictive model with a minimum error.

Algorithm 1. exponential smoothing method

Input: V: vector contains data about production.
Output: alpha, beta: real, holt's method parameters.
begin
// we subdevise V to two parts (testing and learning)
1) A <- Vlearn ; // A contains the vectors for learning
2) T <- Vtest ; // T contains the vectors for testing
3) Generate (alpha, beta) ; // alpha and beta take the maximal values close to 1
4) Xh <- Calculate (V, alpha, beta) ; //calculate the prediction (Xh)
5) E1 <- Error (Xh,T) ; //calculate the error
6) **repeat**
7) Modified (alpha, beta) ;
8) Xh <- Calculate (V, alpha, beta) ;
9) E <- Error (Xh,T) ; //calculate the new error
10) **If** (E < E1) **then**
11) Save (alpha, beta) ; // save alpha and beta
12) Swap(E, E1) ; // save the minimal error
13) **endif**
14) **until** (steady_State) ;
15) **end.**

2. **Supports vectors machines (SVM)**

In the literature, we ca find two types of SVM method the first one for classification and the other for regression [13]. Since we are interested in production that needs real values, in this paper, we have used SVM for regression [14]. The support vector machine SVM is a very effective method for Machine Learning; it is based on the statistical theory of the learning. In addition, this method avoids the traditional process of induction of the deduction, to simplify the problems of classification and regression problems.

The Principle of SVMs. SVM method is based on the first two steps learning and the second is to use the model. Before starting the algorithm, we need to define the used data in this method, which are the label and features. In our case, the label represents the production value and the features are the climatic factors. These steps illustrate as follows:

- Training and model validation phase. Training follows the principle of the diagram shown in the next Fig. 5.

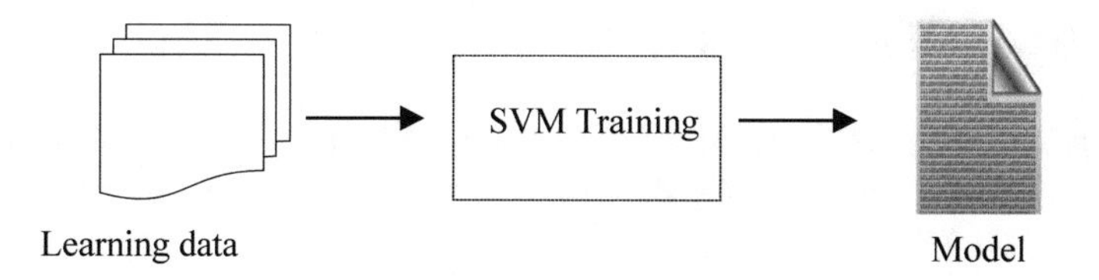

Fig. 5. SVMs training step.

To validate the model obtained we need to test several times based on a test in order to minimize the error we have used the root mean square error (RMSE) [15]:

$$\mathrm{Error} = \sqrt{1/N \sum\nolimits_{t=1}^{N} \left[X_t - \hat{X}_t\right]^2} \tag{6}$$

– Test model step is given in Fig. 6.

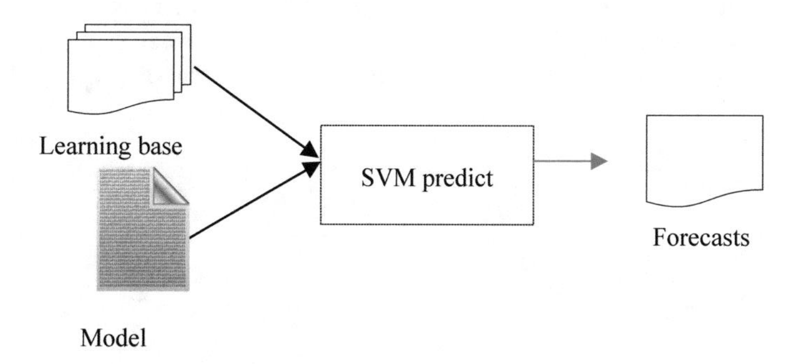

Fig. 6. SVMs test model step.

– To use the obtained forecasts we need to set the climatic factors. Figure 7 shows the using model step.

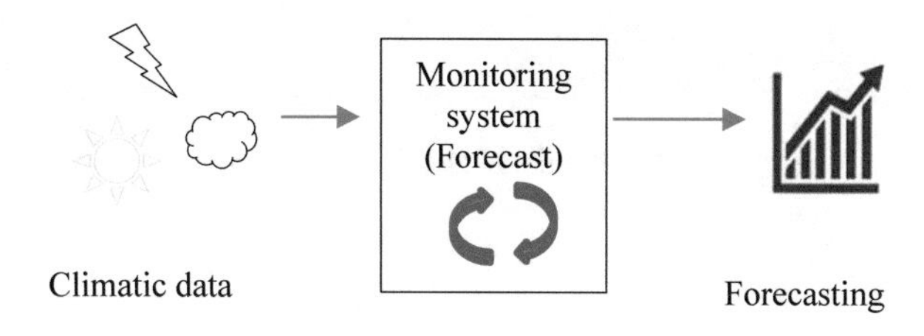

Fig. 7. Using model step.

6 Implementation and Results

In this section, we present the simulation settings that help us to implement our proposed architecture. Furthermore, this section presents some results and introduces graphical interfaces.

6.1 Simulation Settings

All simulation results were measured with a Sun Java SE 7 VM running on Windows 7 PC with an Intel processor with 3.2 GHz and 8 GB memory space.

In our experiment, we have used production data between 1990 and 2016 years, as we mentioned earlier we used the number of palms the productive ones since these are many types of them. We have used four climatic factors: temperature, humidity, wind speed and precipitation. To simulate the number of palms we adjust 10000 palms tree in order to create the clusters using k-means.

6.2 Obtained Results

In this subsection, we going to presents some interfaces performed by our system presented using graphical schemes for the two cases statistical research about the production per year (in general or detailed way), or some forecasts (Figs. 8, 9 and 10).

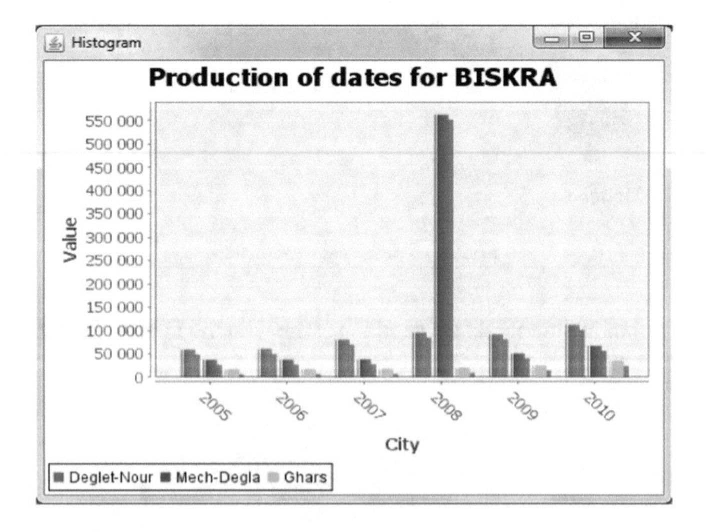

Fig. 8. Search results as a histogram.

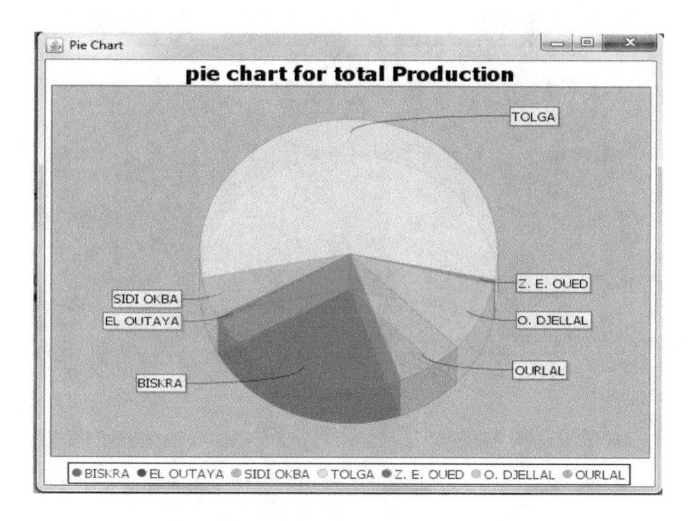

Fig. 9. Search results as a pie chart.

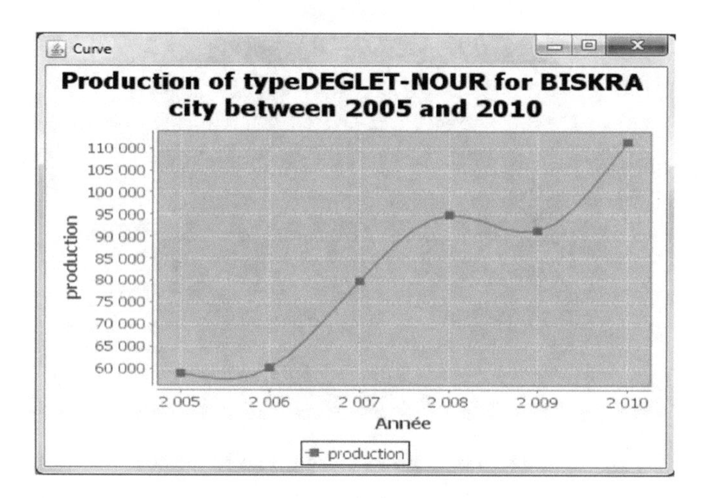

Fig. 10. Search results as a curve.

– The obtained results of the statistical search are:

The Next figures show the results of forecasting during model construction (Training step):

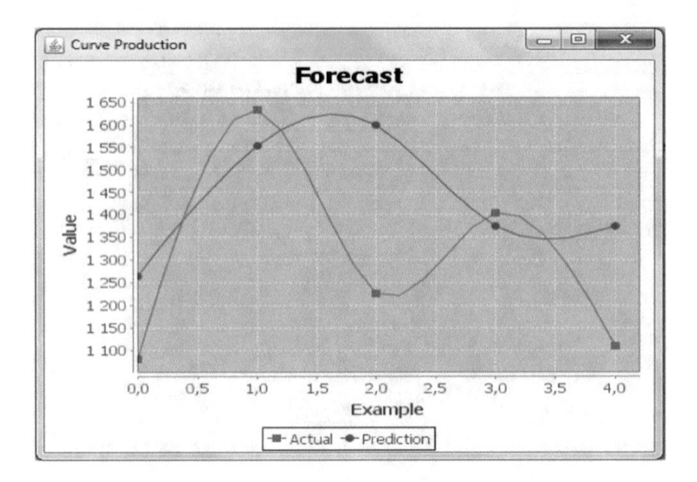

Fig. 11. Training step using the exponential smoothing method.

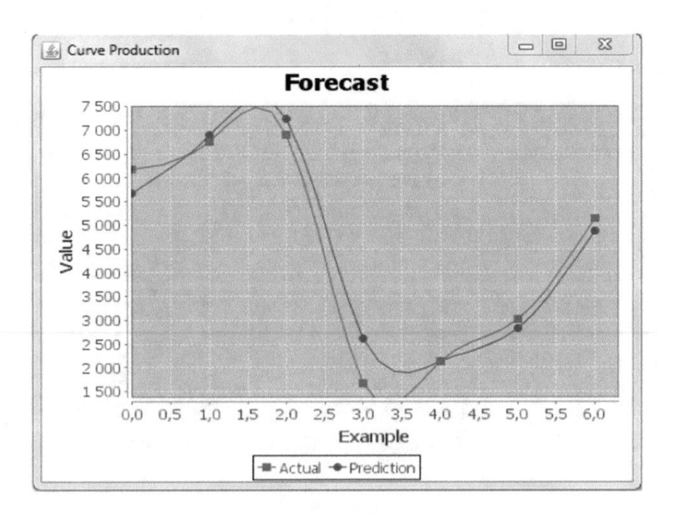

Fig. 12. Training step using the SVM method.

From the previous graph presented in Figs. 11 and 12, we can see that using the climatic factors could enhance the forecast compared to the Holt's exponential smoothing method. The following Table 1 shows a comparison between the two methods of forecasting:

Table 1. Comparison between the two forecasting used methods.

Holt's exponential smoothing	SVM for regression
Number of parameters = 2	Number of parameters > 2
Works only on digital data	Works on any type of data
It only works on production data	Works on the production and climatic factors influencing
It uses real data	It uses normalized data

7 Conclusion

The aim of the presented work is to develop software capable of monitoring production. Moreover, it gives forecasts of dates production to help the data analysts in decision making process. Our system is designed to give to kind of results either simple users through general knowledge about the dates fruits data in some region (as an information system) or decision-maker through statistical results (graphics scheme or forecasts). To achieve this goal, we used two methods of forecasting, the first is based only on the history of the production and second one is based on climatic factors that influence the production in order to achieve the main goal is to improve total production in Algeria. As we have seen this paper, intelligent technologies can help to improve results using drones and sensors in collecting data process.

As we have seen in forecasts method, in exponential smoothing we found that this method is based on production only, and the results may take a linear form (due to the prediction formula: "ax + b"), results may give poor results because they do not consider the causes that influence the production. As against, the previous method SVR method takes into consideration the features that influence production, but they can take bad results due to the number of instances and the number of features.

In the future work, we plan to add a solution for the drone navigation path in order to reduce the navigated distance and the energy consumed during the process.

References

1. Bolman, B., Jak, R.G., van Hoof, L.: Unravelling the myth–the use of decisions support systems in marine management. Mar. Policy **87**, 241–249 (2018)
2. Bhardwaj, R.K., Bhardwaj, V., Singh, D.P., Gautam, S.S., Saxena, R.R.: Modelling and forecasting of wheat production through structural time-series models in Chhattisgarh. Int. J. Pure App. Biosci. **5**(5), 212–216 (2017)
3. Deshmukh, S.S., Paramasivam, R.: Forecasting of milk production in India with ARIMA and VAR time series models. Asian J. Dairy Food Res. **35**(1), 17–22 (2016)
4. Amin, M., Amanullah, M., Akbar, A.: Time series modeling for forecasting wheat production of Pakistan. Plant Sci. **24**(5), 1444–1451 (2014)
5. Lakshmanan, D., Meeran, A.N.: NFC logging mechanism—forensic analysis of NFC artefacts on android devices. In: Artificial Intelligence and Evolutionary Computations in Engineering Systems, pp. 93–101. Springer, Singapore (2017)
6. Floreano, D., Wood, R.J.: Science, technology and the future of small autonomous drones. Nature **521**(7553), 460 (2015)
7. Yurish, S.Y., Gomes, M.T.: Smart sensors and MEMS: Proceedings of the NATO Advanced Study Institute on Smart Sensors and MEMS, Povoa de Varzim, Portugal, 8–19 September 2003, vol. 181. Springer Science & Business Media (2005)
8. Zouai, M., Kazar, O., Haba, B., Saouli, H., Benfenati, H.: IoT approach using multi-agent system for ambient intelligence. Int. J. Softw. Eng. Appl. **11**(9), 15–32 (2017)
9. Cheung, W., Leung, L.C., Tam, P.C.F.: An intelligent decision support system for service network planning. Decis. Support Syst. **39**, 415–428 (2005)
10. Wu, J.: Advances in K-means Clustering: a Data Mining Thinking. Springer, Heidelberg (2012)

11. Aloui, I., Kazar, O., Kahloul, L., Servigne, S.: A new itinerary planning approach among multiple mobile agents in wireless sensor networks (WSN) to reduce energy consumption. Int. J. Commun. Netw. Inf. Secur. (IJCNIS) **7**(2), 116–122 (2015)
12. Bourbonnais, R., Michel, T.: Analyse des séries temporelles: applications à l'économie et à la gestion, edt. Dunod, Paris (2004)
13. Hamza, S., Abderaouf, G., Abdelhak, M., Okba, K.: A new cloud computing approach based SVM for relevant data extraction. In: Proceedings of the 2nd International Conference on Big Data, Cloud and Applications, p. 1. ACM, March 2017
14. Merizig, A., Kazar, O., Lopez-Sanchez, M.: A dynamic and adaptable service composition architecture in the cloud based on a multi-agent system. Int. J. Inf. Technol. Web Eng. (IJITWE) **13**(1), 50–68 (2018)
15. Yang, T., Cuixia, L.: The study on livestock production prediction in heilongjiang province based on support vector machine. In: Proceedings of the 2nd International Conference on Computer Science and Electronics Engineering (ICCSEE 2013), Published by Atlantis Press, Paris, France
16. Chai, T., Draxler, R.R.: Root mean square error (RMSE) or mean absolute error (MAE)?–Arguments against avoiding RMSE in the literature. Geoscientific Model Dev. **7**(3), 1247–1250 (2014)

Impacts of Climate Change on the Production, Yield and Cost of Adaptation of Varieties Imported from Strawberry Plants in the Perimeter of Loukkos (Morocco)

Mohammed Ezziyyani[1(✉)], Ahlem Hamdache[2], Mostafa Ezziyyani[3], and Loubna Cherrat[4]

[1] Polydisciplinary Faculty of Larache, Department of Life Sciences, Abdelmalek Essaâdi University, B.P.: 745 Poste Principale, 92004 Larache, Morocco
mohammed.ezziyyani@gmail.com

[2] Faculty of Sciences of Tetouan, Department of Biology, Abdelmalek Essaâdi University, Avenue de Sebta, Mhannech II, 93002 Tetouan, Morocco

[3] Faculty of Sciences and Techniques, Department of Computer Sciences, Abdelmalek Essaâdi University, Ancienne Route de l'Aéroport, Km 10, Ziaten, B.P.: 416, Tangier, Morocco

[4] Faculty of Sciences, Department of Computer Sciences, Chouaib Doukkali University, Avenue des Facultes, B.P.: 20, 24.000, El Jadida, Morocco

Abstract. On a technical level, the cultivation of strawberry in Morocco has developed remarkably during the last 20 years. During the 2016–2017 crop year, this crop covers 3.050 hectares of land, including 180.378.742 strawberry plants imported from various varieties: Sabrina, San Andreas, Fortuna, Festival, Camarosa, Splendor and others. The period from 1990 to 2010, the dominant varieties that were grown are Chandler, OsoGrande and especially Camarosa and this thanks to its very high productivity, profitability, precocity, quality and adaptation to agroclimatic conditions of the perimeter of Luokkos. Moreover, from 2010, the Californian variety Camarosa (and others) experienced a dramatic decline!. The 2009 fall season is considered the hottest, and 2009 is the second warmest of the decade after 2005. From September to November 2009, maximum temperatures were very high during the day, and very low temperatures at night; hence the thermal gap and therefore new unusual meteorological weather events. Farmers had lost patience because of low yields (very low productivity <500 g/plant) and doubts were starting about the choice (s) of the variety (s)!. A decline of the varieties Camarosa, Festival, Splendor and a total disappearance of varieties Amiga and Benicia. This upset the choice of the distribution of varieties of strawberry plants imported in 2017. Today, many varieties are disappearing Moroccan producers, the choice being dictated by the production objectives. It must be emphasized that the development of a new variety of native strawberries is more important than ever; knowing that it is a work of long breath.

M. Ezziyyani (Ed.): AI2SD 2018, AISC 911, pp. 37–45, 2019.
https://doi.org/10.1007/978-3-030-11878-5_4

Keywords: Climate change · Strawberry · Interannual variability · Loukkous

1 Introduction

Strawberries have been grown and foraged for millennia, although the modern *Fragaria* × *ananassa* was not developed until the 18th century [1]. Native peoples of North America gathered *Fragaria virginiana*, or *Virginia* strawberries, with the berries making appearances in Cherokee myths. The Ancient Romans and Greeks grew *Fragaria vesca*, the alpine strawberry native to Europe and North America, for medicinal and landscape purposes initially, later for consumption [2]. In medieval times, Europeans also cultivated native *Fragaria moschata*, a muskier flavored berry and the green *Fragaria viridis* for ornamental purposes. *F.vesca* dominated cultivation in Europe until the 17th century (with a brief dip in popularity after the 12th century declaration from the abbess Saint Hildegard von Binger that strawberries were not to be eaten, as they grew on the ground among snakes and toads) when *F. virginiana*, recently imported from North America, began to replace it [2]. In 1716, French spy Amédée Frézier brought the South American *Fragaria chilonensis*, long cultivated by the Mapuche people of Chile, to France. In the 1760, Nicholas Duchesne, a French botanist, began noticing 'unusual' strawberry seedlings with larger, redder fruit and a sweet smell in gardens in Brittany. By 1766, Duchesne determined that this new cultivar was a cross between *F. virginiana* and *F. chilonensis*. He named the new interspecific hybrid *F.* × *ananassa* after the sweet pineapple (*anana* in French) aroma; they spread throughout France and eventually the world. *Fragaria vesca* is still grown as a delicacy in some home gardens, while *F. chilonensis* can still be found in parts of Chile, and *Fragaria moschata* and *F. viridis* have declined significantly. All have been replaced by the sweeter, more popular *F.* × *ananassa* cultivars. The strawberry (*Fragaria spp.*) is in the Rosaceae, along with apples, plums, and several other major fruits. It is a perennial, herbaceous, and diminutive plant, growing low to the ground 3 from a central crown. Roots reach about 15–30 cm into the ground below the crown and live for several years. Most cultivars have trifoliate, compound leaves arranged in a spiral around the crown. Strawberries reproduce sexually via seed; they can also reproduce asexually using their stolons, or runners, which grow several cm away from the 'mother' plant before rooting into the ground at the nodes and developing a new crown. Several weeks later, the stolon deteriorates and the new 'daughter' clone is independent from its 'mother', eventually growing its own flowers, leaves, and stolon [3, 4]. The strawberry 'fruit' is not actually a berry but instead a receptacle for potentially hundreds of tiny fruits, known as achenes. Flowers have five or more white obovate petals that surround up to thirty stamens, which in turn surround a raised, conical receptacle covered in up to 500 pistils, each atop an individual carpel [1]. After pollination occurs, this receptacle enlarges and turns red, developing into the 'berry' that we eat. The actual fruits are achenes that develop from each carpel, holding one seed each on the receptacle's surface [4]. Wild strawberries or native species in the genus Fragaria can be found on four different continents. Lake Baikal, in Siberia, separates the nearly one dozen Asian cultivars from the two native European species, with at least two cultivars in North America and one in Chile. The modern, commercially available strawberry is a hybrid of an American and a Chilean wild strawberry, *Fragaria virginiana* and *F. chiloensis*, respectively, and is fully described as *Fragaria* × *ananassa* Duchesne ex Rozier nothosubsp. Ananassa [1]. Different cultivars of *F.* × *ananassa* have been developed for growth in many different climates —anywhere from the taiga to the subtropics. Most commercial production occurs in temperate and Mediterranean climates, between the 42nd north and south parallels. Strawberry breeding programs can be found all across the world, but most are centered in the United States and Europe. The University of California and the University of Florida are two large public sector breeding programs in the United States [5, 6]. Whitaker et al. 2011; another is the USDA Horticultural Crops Research Lab in Oregon, which developed the popular 'Hood' cultivar [5] as well as the University of Minnesota which developed 'Mesabi' and 'Winona' [7]. There are several public breeding programs in Europe as well, such as Wageningen University in The Netherlands, which produced the still popular 'Elsanta'. Private breeding programs expanded worldwide in the 1980s and 1990s, with California's Driscoll Strawberry Associates, a private sector breeder and Grower Company, dominating in the United States. Other large European private breeders include Centro Innovazione Varietale in Italy and Planasa in Spain [5]. The red berries sector, including Strawberry, Raspberry and Blueberry crops, is in its third decade of development since its birth in the perimeters of Loukkos and Gharb in North West Morocco. This development has been favored by the proximity of Europe, the favorable climate, the availability of land, water and skilled labor. This sector and particularly strawberry knows constraints inherent to its profitability which are invented as follows:

- The choice of plant material: problems of varieties, technical choices and commercial problems.
- The profitability of this crop is limited to the level of the small farm.
- The technical supervision at the level of the technical routes and health supervision.

2 Materials and Methods

2.1 Plant Materials

The genus *Fragaria* includes 20 wild species of different ploidy levels, from diploid (e.g. *F. vesca* woody strawberry, 2n = 2x = 14) to octoploid (2n = 8x = 56) (Fig. x). A decaploid species, *F. iturupensis*, has very recently been described. Of the 20 strawberry species, 11 are diploid, five tetraploid, one hexaploid, two octoploid and one decaploid. The cultivated strawberry, *F. x ananassa*, is an interspecific hybrid resulting from the cross between two octoploid species *F. virginiana* and *F. chiloensis*. In the genus *Fragaria*, the polyploid appears to result from the unification of unreduced gametes. This event could be of the order of 1%.

2.2 Temperature Tests

There are two distinct terms called "prediction" and "projection" often used the wrong way when talking about simulations of future climate. The term "projection" is used in general as description of the future and the pathway leading to it. More specifically the Intergovernmental Panel on Climate Change (IPCC) uses the term "climate projection" when referring to model-derived estimates of future climate. NASA Goddard Institute for Space Studies and Center for Climate Systems Research, provided in 2008 following definitions: Predictions are estimated outcomes under highly specific conditions and imply certainty; they are not exclusively restricted to the future, while projections are predictions conditional on a future scenario. Projections are describing a range of possible developments, some for near, some for remote future. Nevertheless they can't be excluded from the list of possible developments (Fig. 1).

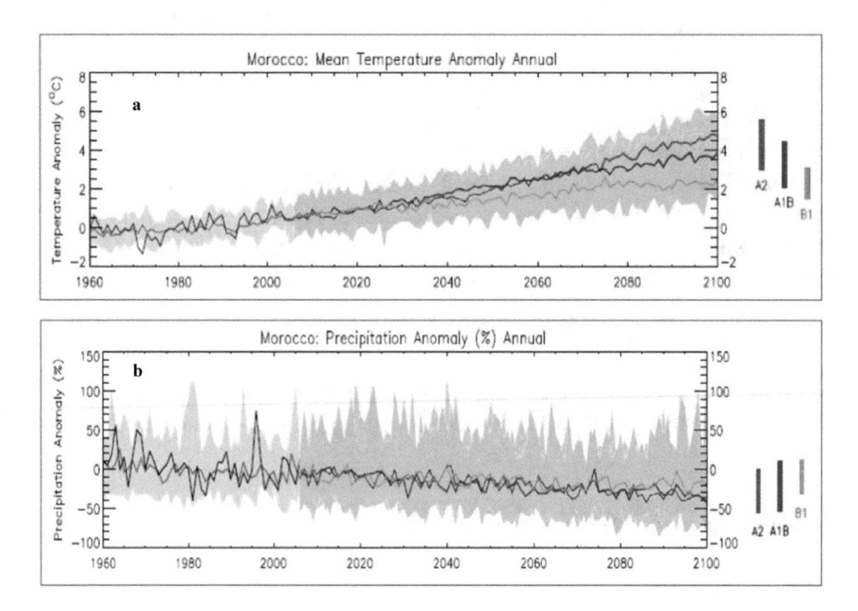

Fig. 1. Trend in annual mean temperature (Fig. a) and precipitation (Fig. b) for the recent past and projected future. Values are anomalies relative to 1970–1999 mean climate. Black curves represent the mean of observed data from 1960 to 2006. Brown colours stand for median (solid line) and range (shading) of model simulations across an ensemble of 15 models. Coloured lines from 2006 represent median and range of the ensemble projections under three emissions scenarios. Coloured bars on the right side summarise the range of mean 2090-2100 climates, again calculated by 15 models for each scenario (Source: C. Mc Sweeney, M. New and G. Lizcano, 2008: Morocco (UNDP Climate Change Country Profiles)).

2.3 Results and Discussion

In terms of the choice of plant material, it can be seen that since the period from 1990 to 2010, only dominant varieties have been cultivated are Chandler Osogrande and

Camarosa. These varieties and especially the Camarosa that lasted from 1990 to 2010 were homogeneous varieties, regular and have shown profitability both productive and financial. For farmers, they were at the base of the growth of this crop at its start. Camarosa easy technical route associated with methyl bromide disinfection has, during this period, provided high yields that ensure high profitability for farmers. Since 2010, we are witnessing the decline of Camarosa (Fig. 2) and its replacement by new varieties

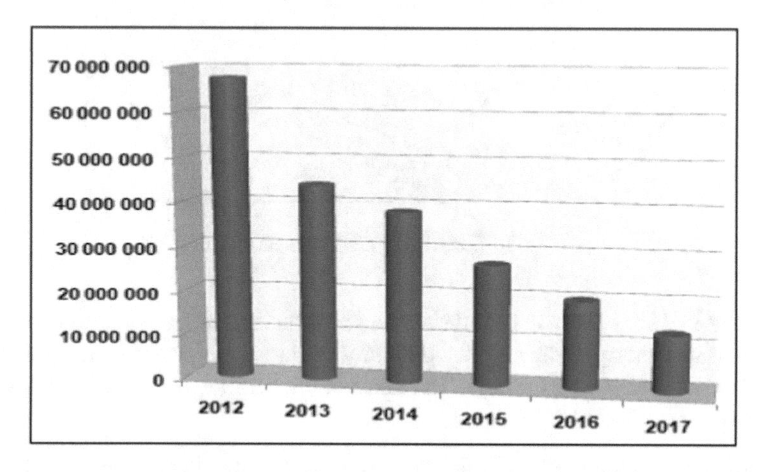

Fig. 2. Evolution of the Camarosa variety

During flowering, the floral transition is a key, highly controlled event that requires the integration of different environmental cues with endogenous physiological markers [8]. In cultivated plants, this key trait is selected to produce varieties better adapted to their environment. This involves selecting varieties with a high degree of plasticity that gives them the ability to adapt to climate change, for example [9]. Flowering in strawberry is governed by the close interaction between photoperiod and temperature [10].

The success of sexual reproduction, corresponding to the development of flowers and fruits, depends on flowering at the right moment, that is to say, at the moment favoring pollination between individuals of the same species and conditions favoring fruit and seed development [11, 12].

Based on the work carried out by the National Meteorological Directorate (NMD) on climate monitoring and detection [13], several significant trends have been identified for both thermal and rainfall parameters. A significant downward trend in rainfall accumulations during the rainy season was observed over the period 1961–2005. It is estimated at about one (–2.6 mm/year). Over the 45 years considered, this decline averages around 26% nationally; the end of the rainy season from February to April shows more important trends, and this coincides exactly with Camarosa variety, and this coincides exactly with the inflorescence emergence of the Camarosa variety (Fig. 3).

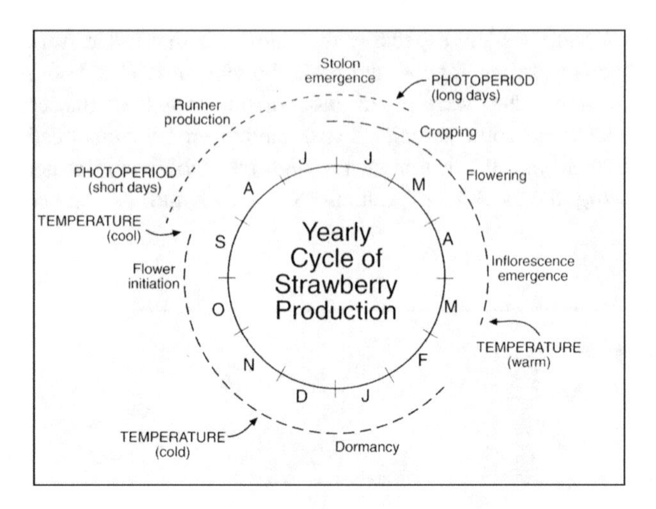

Fig. 3. Strawberry development cycle

Already in 1966, Darrow [14] described the influence of temperature on the floral initiation of strawberry, indicating the difficulty of classifying varieties according to the photoperiod as is generally the case. The effect of temperature on strawberry bloom can be summarized as follows:

- for low temperatures (<9 °C), the flowering of the strawberry is inhibited,
- between 10 °C and 15 °C, flowering is activated whatever the photoperiod conditions [15, 16],
- between 15 °C and 25 °C, the genotype flowers according to photoperiod conditions,
- above 25 °C, strawberry flowering is inhibited [17]. Responses are similar in diploid and octoploid strawberries, although temperature optima and limits are often higher for *F x. ananassa* that for *F. vesca* [10, 18–20], Cold requirements are required to remove bud dormancy and allow further floral development in the spring [21]. Cold requirements are highly dependent on genotype [22]. They can be defined as the number of cumulative hours between 0 °C and 7 °C to satisfy the needs of strawberry, that is to say allowing a resumption of normal growth when temperatures are rising. Vernalization promotes vegetative and reproductive development [22]. Vernalization promotes vegetative growth and reduces floral induction, which promotes floral differentiation [23]. A non-vernalized genotype produces smaller fruits [24] and lower quality [25] than the same genotype with adequate vernalization. Vernalization is the process by which the floral transition is activated by prolonged exposure to cold. The need for vernalization ensures flowering in conditions that should be favorable to the reproductive phase. Vernalization is a quantitative process. Indeed, more exposure to cold is longer and the flowering is accelerated. The perception of vernalization is at the apical vegetative meristem [26]. Unlike the photoperiod, vernalization prepares the plant to flower but can not by itself induce flowering. There is a temporal separation between the

cold treatment and the flowering. After a more or less prolonged exposure to cold, the plant resumes vegetative growth for several weeks with warmer temperatures, then flowering will be initiated. This procedure assumes that the perception of low temperatures must be memorized at the apex and during mitosis [27]. This vernalized state is however lost after meiosis [12]. A process for stable memorization across cell divisions suggests that the vernalization response involves epigenetic regulation [27].

3 Conclusion

The floral transition is a key event in plant life. In strawberry, understanding the genetic mechanisms of floral transition is a major issue for better control of yield fruit production and understands the effect of climate change on the adaptation of varieties imported from strawberries and to decipher the genetic and molecular mechanism of the continuous flowering in *Fragaria.* As at the global and regional scale, Morocco is not spared from the warming situation. By being part of the Mediterranean basin and the African continent, which are among the most vulnerable to climate change, Morocco is already suffering the negative effects of this change. The fourth report of the Intergovernmental Panel on Climate Change is very clear in its conclusions describing the vulnerability of the African continent, which is also of interest to Morocco; given its central geographic position in population movements related to climate change. In addition, it is useful to note the absence of natural varieties adapted to local production conditions and to our markets, which last from November to March of each year.

Flowering marks the transition between the vegetative state and the reproductive state of a plant. It is an essential step in the process of sexual reproduction in flowering plants or angiosperms by the establishment of reproductive organs of the plant. Flowering would be a biological advantage given the prevalence of angiosperms is the largest and most diverse group of the plant kingdom. The success of sexual reproduction, corresponding to the development of flowers and fruits, depends on flowering at the right moment, that is to say, at the moment favoring pollination between individuals of the same species and conditions favoring fruit and seed development. During flowering, the floral transition is a key, highly controlled event that requires the integration of different environmental cues with endogenous physiological markers. In cultivated plants, this key trait is selected to produce varieties better adapted to their environment. The aim is to select varieties with a high degree of plasticity, providing them with a capacity to adapt to climate change, for example.

In fact, the expectations of strawberry producers for varietal innovation are rather diffuse and complex. It is thus the choice of a variety adapted to the edapho-climatic conditions of the Loukkous perimeter and the quality of the starting plant constitute the first step towards a good production and yield of strawberry.

Morocco is already feeling the brunt of climate change after the severe drought of 2015 that decimated crops and depressed the economy. Morocco has adopted the "Green Morocco Plan" to face the threat of climate change, and aims to make agriculture one of the engines of growth of the national economy.

References

1. Galletta, G.J., Himelrick, D.C. (ed.): Currant and gooseberry management—cultivars. In: Small Fruit Crop Management, United States, vol. 602, pp. 254–258. Prentice Hall, Englewood Cliffs (1989)
2. Hummer, K., Hancock, J.F.: Strawberry genomics: botanical history, cultivation, traditional breeding, and new technologies. In: Folta, K., Gardiner, S. (eds.) Genetics and Genomics of Rosaceae, pp. 413–436. Springer, New York (2009)
3. Martin, E., Tepe, E. (eds.): Cold Climate Strawberry Farming. Department of Horticultural Science e-book, University of Minnesota, St. Paul, MN (2014)
4. Davis, T.M.: Geographical distribution of strawberries. University of New Hampshire, New Hampshire Agricultural Experiment Station, Durham, NH (2015). http://strawberrygenes.unh.edu/map.html. Accessed 2 Aug 2018
5. Faedi, W., Mourgues, F., Rosati, C.: Strawberry breeding and varieties: situation and perspectives. In: ISHS Acta Horticulturae 567: IV International Strawberry Symposium (2002)
6. Whitaker, V.M., Hasings, T., Chandler, C.K., Plotto, A., Baldwin, E.: Historical trends in strawberry fruit quality revealed by a trial of University of Florida cultivars and advanced selections. HortScience **46**, 553–557 (2011)
7. Hoover, E.H., Tepe, E.S., Foulk, D.: Growing strawberries in Minnesotagardens (2015). http://www.extension.umn.edu/garden/yard-garden/fruit/strawberries-for-thehome-garden/#cultivars. Accessed 2 Aug 2018
8. Tremblay, R., Colasanti, J.: Floral Induction. Blackwell Publishing Ltd, Oxford (2007)
9. Jung, C., Muller, A.E.: Flowering time control and applications in plant breeding. Trends Plant Sci. **14**, 563–573 (2009)
10. Bradford, E., Hancock, J.F., Warner, R.M.: Interactions of temperature and photoperiod determine expression of repeat flowering in strawberry. J. Am. Soc. Hortic. Sci. **135**, 102–107 (2010)
11. Simpson, G.G., Dean, C.: Flowering - Arabidopsis, the rosetta stone of flowering time? Science **296**, 285–289 (2002)
12. Putterill, J., Laurie, R., Macknight, R.: It's time to flower: the genetic control of flowering time. BioEssays **26**, 363–373 (2004)
13. Driouech, F.: Etude des indices de changements climatiques sur le Maroc: températures et précipitations. INFOMET? Casablanca, November 2006 (2006)
14. Darrow, G.M.: The Strawberry. History, Breeding and Physiology. Holt, Rinehart and Winston, New York (1966)
15. Brown, T., Wareing, P.F.: The genetical control of everbearing habit and three other characters in varieties of *Fragaria vesca*. Euphytica **14**, 97–112 (1965)
16. Guttridge, C.G.: *Fragaria* x *ananassa*. CRC Handbook of Flowering III, pp. 16–33 (1985)
17. Ito, H., Saito, T.: Studies on the flower formation in the strawberry plant I. Effects of temperature on flower formation. Tohoku J. Agric. Res. **13**, 191–203 (1962)
18. Sonsteby, A., Heide, O.M.: Long-day control of flowering in everbearing strawberries. J. Hortic. Sci. Biotechnol. **82**, 875–884 (2007)
19. Sonsteby, A., Heide, O.M.: Quantitative long-day flowering response in the perpetual-flowering F-1 strawberry cultivar Elan. J. Hortic. Sci. Biotechnol. **82**, 266–274 (2007)
20. Sonsteby, A., Heide, O.M.: Long-day rather than autonomous control of flowering in the diploid everbearing strawberry *Fragaria vesca* ssp semperflorens. J. Hortic. Sci. Biotechnol. **83**, 360–366 (2008)

21. Stewart, P.J., Folta, K.M.: A review of photoperiodic flowering research in strawberry (*Fragaria* spp.). Crit. Rev. Plant Sci. **29**, 1–13 (2010)
22. Darnell, R.L., Hancock, J.F.: Balancing vegetative and reproductive growth in strawberry. In: Chandler, C.K., Pritts, M.V., Crocker, T.E. (eds.) The IV North American Strawberry Conference (Hort. Sci. Dept., Univ. of Florida, Gainesville, FL), pp. 144–150 (1996)
23. Durner, E.F., Poling, E.B.: Flower bud induction, initiation, differentiation and development in the Earliglow strawberry. Sci. Hortic. **31**, 61–69 (1987)
24. Hamann, K.K., Poling, E.B.: The influence of runner order, night temperature and chilling cycle on the earliness of 'Selva' plug plant fruit production. Acta Hort. **439**, 597–603 (1997)
25. Bringhurst, R.S., Voth, V., Van Hook, D.: Relationship of root starch content and chilling history to performance of California strawberries. Proc. Am. Soc. Hortic. Sci. **75**, 373–381 (1960)
26. Boss, P.K., Bastow, R.M., Mylne, J.S., Dean, C.: Multiple pathways in the decision to flower: enabling, promoting, and resetting. Plant Cell **16**, 18–31 (2004)
27. Henderson, I.R., Shindo, C., Dean, C.: The need for winter in the switch to flowering. Annu. Rev. Genet. **37**, 371–392 (2003)

Bacillus amyloliquefaciens Enhanced Strawberry Plants Defense Responses, upon Challenge with Botrytis cinerea

Meryem Asraoui[1(✉)], Filipo Zanella[2], Stefania Marcato[2], Andrea Squartini[3], Jamila Amzil[1], Ahlem Hamdache[4], Barbara Baldan[2], and Mohammed Ezziyyani[5]

[1] Faculty of Sciences and Techniques of Settat, University of Hassan 1er, Settat, Morocco
meryasra@gmail.com
[2] Department of Biology, University of Padua, Padua, Italy
[3] Department of Agronomy Food Natural Resources Animals and Environment, Legnaro, PD, Italy
[4] Department of Biology, Faculty of Sciences of Tetouan, University of Abdelmalek Essaâdi, Tetouan, Morocco
[5] Department of Life Sciences, Polydisciplinary Faculty of Larache, University of Abdelmalek Essaâdi, Larache, Morocco
mohammed.ezziyyani@gmail.com

Abstract. Plants activate a range of active defense strategies in response to biotic stresses. Systemic acquired resistance (SAR) occurs when the plants are triggered by a stimulus prior to infection by a plant pathogen able to reduce the severity of the disease. The most widely studied group of PGPB is plant growth-promoting rhizobacteria that colonize the root surface. This trait plays a crucial role in protecting plants from pathogen attack since poor colonization could cause decreased antagonistic activity. The present work aims to isolate, identify and select strains from the rhizosphere of strawberry in the Larache area, with potential application as a biocontrol agent against gray mold disease caused by *Botrytis cinerea*. First, the selected bacteria and the pathogen were characterized using 16S rDNA and ITS sequencing respectively, and then challenged in dual culture as antagonists to evaluate their effect on the growth of *B. cinerea*. Among the selected bacteria that inhibited the growth of *B. cinerea* in vitro dual culture, *Bacillus amyloliquefaciens* was able to associate to the external root surfaces, by forming biofilm, moreover, preliminary analyses showed that, upon attack with Botrytis cinerea, B. amyloliquefaciens primed strawberry's plants, showed an enhanced expression of PR1 and β-1,3-glucanase (FaBG2-2) genes in comparison to non-primed plants. However, no expression of chitinase II genes (FaChi2-1, FaChi2-2) was observed. This "priming" effect indicated that B. amyloliquefaciens could be further studied for its biocontrol traits able to induce, once inoculated, defense-related gene expression earlier and stronger than in non-primed plants.

Keywords: *Botrytis cinerea* · *Bacillus amyloliquefaciens* · *Fragaria x ananassa Duch.* · Biological control · Defense-related gene

M. Ezziyyani (Ed.): AI2SD 2018, AISC 911, pp. 46–53, 2019.
https://doi.org/10.1007/978-3-030-11878-5_5

1 Introduction

Plant growth promoting Rhizobacteria (PGPR) is a group of beneficial free-living soil bacteria that can be found in the rhizosphere soil. PGPRs are able to colonize very efficiently the rhizosphere soil of the plants or their rhizoplane (root surface), or the root itself (within radicular tissues) [1, 2]. PGPRs are known for their beneficial effect on the plant by promoting the growth directly via indirect or direct mechanisms or the combination of both mechanisms. The direct promotion of plant growth by PGPR involves their abilities to synthesis different compounds such as plant growth regulators (auxin, gibberellins, ethylene) to modulate root development and growth, or facilitating the nutrients uptakes through nitrogen fixation, solubilization of phosphorus, or ammonium production. The indirect promotion of plant growth occurs when PGPR reduce the impact of diseases through antagonism to soil-borne pathogens by different mechanisms, which include antibiosis, competition for nutrients, synthesis of a wide range of secondary metabolites such as siderophores, extracellular cell wall degrading enzymes, antimicrobial and antifungal compounds and cyclic lipopeptides, or by triggering induced systemic resistance (ISR) or inducing systemic acquired resistance (SAR). The establishment of systemic acquired resistance (SAR) is often associated with the expression of a set of pathogenesis-related protein genes [3], SAR occurs when plants activate their defense mechanism in response to a primary infection by necrotizing pathogens, especially when they induce a hypersensitivity reaction by which it becomes limited in a local necrotic lesion of the brown dried tissue [4]. Some pathogenesis proteins have been identified as β-1,3-glucanases and chitinases, able to hydrolyze fungal cell wall components [3]. Therefore, the accumulation of PR-proteins has often been proposed as the molecular basis of SAR. To exert their beneficial effects in the root environment, rhizobacteria must be able to compete with other microbes in the rhizosphere for nutrients and for root exudates such as amino acids and sugars, secreted by the root to attract specific bacteria on their sites providing them a rich source of energy and allowing root colonization. This root competence plays a crucial role in the protection of plants against pathogenic attacks since poor colonization could lead to effectively decrease antagonistic activity. Furthermore, the combination of the different mechanisms of action used by bacteria to control pathogens might be useful in biocontrol strategies to improve the crop productivity and health. The present work aims to evaluate the effects of *B. amyloliquefaciens* strain in triggering SAR by evaluating the expression of two pathogenesis-related protein genes in *Fragaria x ananassa* plants previously inoculated with the *B. amyloliquefaciens* strain and infected with the fungal pathogen *B. cinerea*, which is responsible for gray mold disease in strawberry plants in some region of Morocco causing a considerable loss in the cultivated fields of the province of Larache (Tangier-Tetouan-Al Hoceima). The biocontrol effect was also evaluated to study the mechanisms responsible for the antagonistic activity of *B. amyloliquefaciens* on the mycelial growth of *B. cinerea* in dual culture as antagonists. In addition, the important role of root colonization in biological control led us to evaluate the bacterial strain's ability to colonize the external root surface.

2 Materiels and Methods

2.1 Vegetal Materiel

Three months old Strawberry (*Fragaria × ananassa, Duch*) plants were used in this work. Plantlets were obtained from the In vitro culture to ensure healthy and bacteria-free plants using the protocol of hamdouni et al. [5].

2.2 Bacterial Inoculation

The strain *B. amyloliquefaciens* (BA) used in inoculation experiments was isolated from the rhizosphere of strawberry's plants. For plant inoculation, bacterial inoculums were prepared on NB medium incubated at 30 °C for 72 h with shaking. Cultures were centrifuged at 6,000 g for 10 min and washed twice with (PBS) to remove any culture medium residue. The inoculation of strawberry plant roots was performed by watering the pots with 5 ml bacterial suspension of about 109 CFU/ml for 28 days prior pathogen infection. This method was used for roots colonization ability and the molecular analysis for gene defense.

2.3 Fungal Inoculation

The strain of *B. cinerea* (Bt.L) was isolated from strawberry's fruit representing symptoms of gray mold disease and grown on PDA medium for 7 days at 28 °C. The mycelia plugs were then placed on the leaves of strawberry's plants for 48 h.

2.4 In Vitro Screening for Antagonism

B. amyloliquefaciens was tested for their ability to inhibit the growth of *B. cinerea* in plates containing PDA medium. The Bacterium was streaked in the center of the plates and mycelial plugs (5 mm) of *B. cinerea* were deposited on the edge of the plates. Plates not inoculated with bacteria were also prepared to serve as a control then incubated at room temperature for 3 days [6].

2.5 Scanning Electron Microscopy

To evaluate the root colonization the Scanning electron microscopy approach was performed using the Environmental Scanning Electron Microscope (ESEM) model FEI Quanta 200F, from the Certification Analysis Center And Services (CEASC), Italy.

2.6 Semi-quantitative RT-PCR

Expression of FaPR1 (PR protein 1), FaBG2-2 (β-1,3-glucanase), FaChi2-1 and FaChi2-2 (class II chitinase) was evaluated from total RNA extracted from leaves of strawberry plants inoculated with *B. amyloliquefaciens* 28 days before challenging with the *B. cinerea* isolate using CTAB 2% protocol. After DNAse I treatment (Promega), 5 μg of total RNA was primed with Random Decamers (Ambion), reverse transcribed with PowerScript Reverse Transcriptase (BD Biosciences CLONTECH), and diluted 1:5. RT-PCR was performed with 5 μL diluted first-strand cDNA, using GAPDH-1 as an internal standard [7, 8]. The thermocycler was programmed with the following parameters: with the following conditions: 10 min at 95 °C, 25 cycles of 30 s at 95 °C, 1 min at the optimal annealing temperature (45–60 °C), and 1 min at 72 °C, followed by 5 min at 72 °C. PCR reactions were allowed to proceed for different number of cycles to determine the exponential phase of amplification. Densitometric analysis of red-stained agarose gels (0.5 μg/mL) was performed using Quantity One software (Bio-Rad). The gene-specific primers used are shown in Table 1. Three independent PCR reactions were performed using the same RNA [7].

Table 1. Primers used in semi-quantitative RT-PCR analysis

Genes GenBank	Accession number	Amplicon size (bp)	Forward primers (5′ to 3′)	Reverse primers (5′ to 3′)
GAPDH-1	AB363963	400	CTACAGCAACACA GAAAACAG	AACTAAGTGCTAAT CCAGCC
FaBG2-2 β-1,3 glucanase	AY989818	350	CTCCATTGTTGCCCAA	AACCCTACTCGGCT GA
FaChi2-1 Chitinase II	AF147091	851	TCGTCACTTGCAA CTCCTAA	GGACTTCTGATTTTC ACAGTCT
FaChi2-2 chitinase II	AF320111	781	CAAGTCAGATAAC AATGGAGAC	TTGTAACAGTCCAA GTGTCC

3 Results and Discussion

The present work is part of a project to characterize and select rhizospheric bacterial strains that could be used as biofertilizers and biological control agents in strawberry's field. The strain *B. amyloliquefaciens* (BA), showing a particular capacity in controlling *B. cinerea* (Bt.L) in vitro by reducing its mycelia growth (Fig. 1), was selected for molecular identification using the 16S rDNA sequencing. After sequencing, the bacterium was compared to other sequences from the gene bank and the analysis of the resulting sequence showed its 99% similarity to Bacillus amyloliquefaciens (NCBI accession number: HQ650780.1). The pathogen used in this study was *B. cinerea* (Bt.L), isolated from strawberry fruit, representing symptoms of gray mold. After macroscopic and microscopic characterization, the molecular identification of the pathogen was performed using ITS sequencing. The results showed 99% similarity with *Botrytis cinerea* (NCBI Accession Number: KY616855.1). The both isolates were selected for further analysis to evaluate their other abilities in biological control.

Using the ESEM approach we confirmed the rhizospheric origin of the *B. amyloliquefaciens* (BA) strain that was able to colonize the root surface of Fragaria × ananassa, Duch forming a sort of biofilm localized around primary roots and in young root zones such as emergent lateral roots (Fig. 2). This root competence can be explained by the positive interaction between the plant and the bacterium and can be due to the production of exo-polysaccharides (EPS) which usually plays an important role in biofilm formation in order to protect the plant against the fungal attack by attachment to surfaces and also in the defense of the plant in microbe-plant interactions [9]. Effective colonization of plant roots by EPS-producing microbes helps to retain free phosphorus in soils and circulate essential nutrients to the plant for proper growth and development [9].

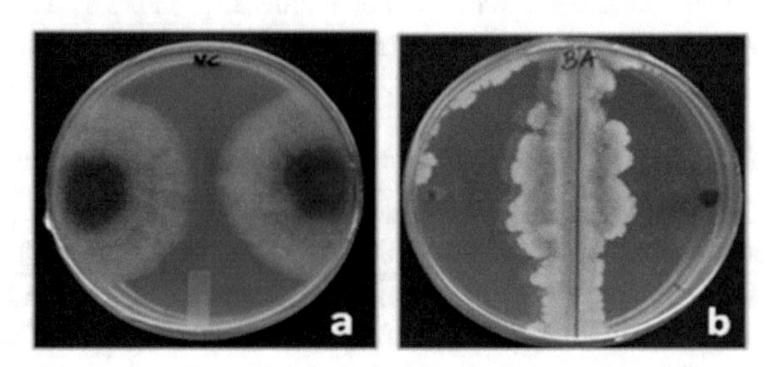

Fig. 1. Antifungal effect of *B. amyloliquefaciens*: a/control (without bacteria) b/reduction of the mycelia growth of *B. cinerea* (Bt.L) in the presence of the *B. amyloliquefaciens* (BA) strain

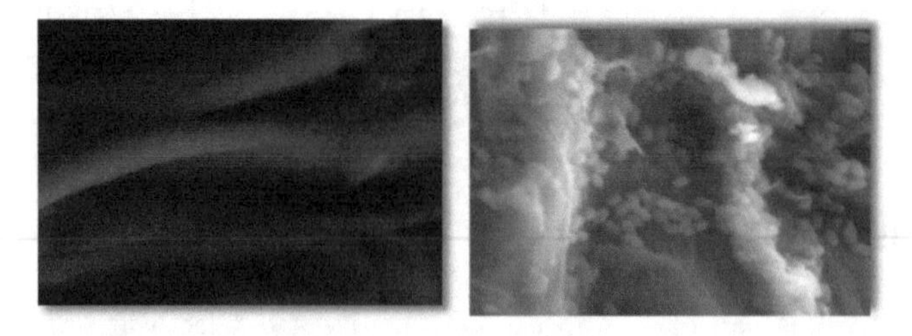

Fig. 2. Rhizospheric colonization of *Fragaria* × *ananassa* roots inoculated with *B. amyloliquefaciens* (BA) observed by scanning electron microscope (ESEM) after 28 days. A/control (without bacteria); B/*Fragaria* × *ananassa* roots inoculated with *B. amyloliquefaciens* (BA) (10.0 μm scale bar)

In addition, the increase of the plant's response to fungal attack is generally associated with the induction of defense-related genes (SAR), so we analyzed the expression of a PR1 protein, two chitinases II and glucanase gene. As shown in Fig. 3, the plants inoculated only with B. amyloliquefaciens (BA) strain caused the induction of defense-related genes such as FaPR1 and FaBG2-2 (Fig. 3b) compared with uninoculated plants (Fig. 3b). When the plants were infected with Botrytis cinerea after bacterial inoculation (Fig. 3c), a strong induction of FaPR1 expression was observed while a slight decrease was observed in the expression of this gene in comparison with plants treated only with *B. amyloliquefaciens* (BA). A strong induction in FaBG2.2 expression was observed in plants treated with *B. amyloliquefaciens* (BA) and infected with B. cinerea (Bt.L) more than the level of expression in plants treated only with the bacterium. For FaChi2-1 and Fachi2.2 no gene expression in all treatments was observed. Confirming the results obtained previously, we observed that the inoculation of *B. amyloliquefaciens* (BA) resulted in an increase in the expression of defense genes compared with uninoculated plants (Fig. 3c).

A/ Development of Systemic Acquired Resistance (SAR) and Activation of the Defense Mechanism in Response to a Primary Infection by B. cinerea (Bt.L)., inducing a Limited hypersensitivity reaction, observed as Local Necrotic Lesion in the leaves.

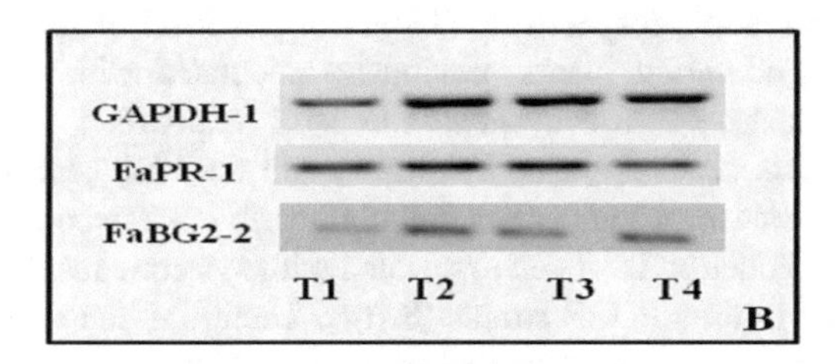

B / Semi-quantitative RT-PCR test for PR1 and BG2.2 genes with 31 amplification cycles. Gels show the reaction products: T1 untreated plants, T2 plants infected with *B. cinerea* (Bt.L), T3 plants inoculated with *B. amyloliquefaciens* (BA), T4 Plants infected with *B. cinerea* (Bt.L). after inoculation with *B. amyloliquefaciens* (BA) .

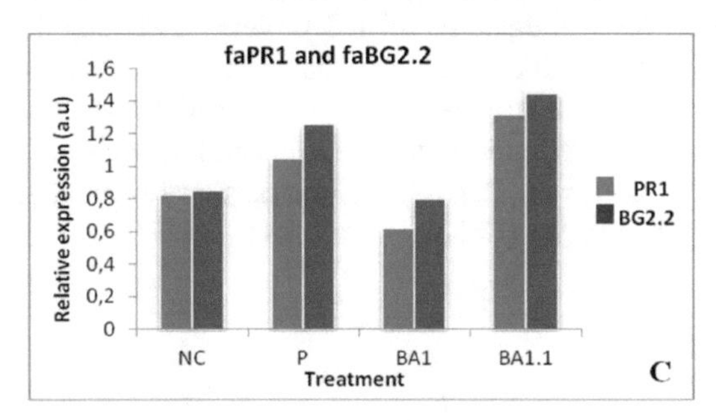

C/ Histogram showing relative expression analysis based on the corresponding gel: each one represents the expression of the normalized gene relative to the expression of GAPDH gene. In order, from left to right: untreated plants, plants infected with Bt.L, plants inoculated with BA, plants infected with Bt.L after inoculation.

Fig. 3. Expression of defense-related genes in strawberry plants before and after inoculation with *B. amyloliquefaciens* (BA) and infection with *B. cinerea* (Bt.L)

The level of expression of the defense genes in plants infected only by the pathogen was not significantly different from the water-treated plants. Considering that the PR1 protein is a molecular marker of the SAR response, these results confirm that *B. amyloliquefaciens* (BA) strain involves the transduction of defense signals prior to pathogenic challenge. Positive expression of glucanase gene (FaBG2-2) was observed only on plants inoculated by *B. amyloliquefaciens* (BA) after infection with *B. cinerea* (Bt.L) while negative regulation of chitinase genes (FaChi2-1 and FaChi2-2) was not observed in the treated plants. However, there are no studies to date on the ability of this bacterium to induce the expression of plant chitinase genes and to indirectly reduce the damage caused by plant pathogens. The glucanases and the chitinases are the most abundant classes of PR-protein genes in strawberry plants with identified hydrolytic activity [10–12]. Previous studies have shown that the genes FaBG2-1 and FaBG2-3 were induced in the leaves of *Fragaria x ananassa* plants after infection with *C. fragariae* or *C. acutatum* [12]. Other studies have also shown that an increase in the total activity of β-1,3-glucanase in strawberry from 2 to 48 h after infection in response to one of the two fungi [11]. As well as FaBG2-1 and FaBG2-3 genes, the FaChi2-1 and FaChi2-2 genes were induced on *C. fragariae* or *C. acutatum* infection in 2–6 or 24–48 h after inoculation, respectively [13]. A higher level of induction was detected when *Fragaria x ananassa* plants previously inoculated with *A. brasilense* REC3 against the anthracnose agent *C. acutatum* M11 expressing the glucanase genes (FaBG2-2), chitinase (FaChi2-1 and FaChi2-2) and the PR1 gene [13]. These results highlight the importance of glucanases and chitinases in strawberry response to biotic stress. Glucanase induction has been associated with systemic resistance against fungal diseases induced by some PGPB strains [8, 14]. Therefore, the accumulation of these hydrolytic enzymes at the fungal hyphal penetration site may result in degradation of the fungal cell walls [14, 15]. In conclusion, we demonstrate that the rhizospheric colonization of strawberries plants with *B. amyloliquefaciens* (BA) strain confers a systemic protection against *B. cinerea* (Bt.L) by the direct activation of certain plant defense reactions like SAR, which increased regulation of defense-related genes, such as those encoding for pathogenesis-related proteins including PR1, chitinases, and glucanase. This “priming” effect indicated that induced plants activate defense-related gene expression earlier and stronger than non-induced plants. Therefore, we propose that the activation of a systemic defense response is associated with the promoting effect of *B. amyloliquefaciens* (BA) strain in strawberry plants against the gray mold disease caused by *B. cinerea* (Bt.L).

4 Conclusion

This study clearly suggests that *B. amyloliquefaciens* could be a potential PGPR that can be used as an effective biocontrol agent for improving the defense and the heath of strawberry plants and controlling gray mold disease caused by *B. cinerea* in Moroccan fields.

References

1. Ahmad, F., Ahmad, I., Khan, M.S.: Screening of free-living rhizospheric bacteria for their multiple plant growth promoting activities. Microbiol. Res. **163**, 173–181 (2008)
2. Gray, E.J., Smith, D.L.: Intracellular and extracellular PGPR: commonalities and distinctions in the plant-bacterium signaling processes. Soil Biol. Biochem. **37**, 395–412 (2005)
3. Saravanakumar, D., Vijayakumar, C., Kumar, N., Samiyappan, R.: PGPR-induced defense responses in the tea plant against blister blight disease. Crop Prot. **26**, 556–565 (2007)
4. Van Loon, L.C., Bakker, P.A.H.M., Pieterse, C.M.J.: Systemic resistance induced by rhizosphere bacteria. Annu. Rev. Phytopathol. **36**, 453–483 (1998)
5. El Hamdouni, E.M., Lamarti, A., Badoc, A.: In vitro germination of the achenes of *Strawberry* (*Fragaria x ananassa Duch.*) *cvs.* "Chandler" and "Tudla". Bull. Soc. Pharm. Bordeaux **140**, 31–42 (2001)
6. Favaro, G., Bogialli, S., Di Gangi, I.M., Nigris, S., Baldan, E., Squartini, A., et al.: Characterization of lipopeptides produced by *Bacillus licheniformis* using liquid chromatography with accurate tandem mass spectrometry. Rapid Commun. Mass Spectrom. **30**, 2237–2252 (2016)
7. Navazio, L., et al.: A diffusible signal from arbuscular mycorrhizal fungi elicits a transient cytosolic calcium elevation in host plant cells. Plant Physiol. **144**, 673–681 (2007)
8. Tortora, M.L., et al.: Protection of strawberry plants (*Fragaria ananassa Duch.*) against anthracnose disease induced by *Azospirillum brasilense*. Plant Soil **356**, 279–290 (2012)
9. Tewari, S., Arora, N.K.: Multifunctional exopolysaccharides from *Pseudomonas aeruginosa* PF23 involved in plant growth stimulation, biocontrol and stress amelioration in sunflower under saline conditions. Curr. Microbiol. **69**, 484–494 (2014)
10. Khan, A.A., Shi, Y., Shih, D.S.: Cloning and partial characterization of a b-1,3-glucanase gene from strawberry. DNA Seq. **14**, 406–412 (2003)
11. Shi, Y.: Isolation, characterization and expression analysis of b-1, 3-glucanase genes from strawberry plants. Thesis, Louisiana State University, Louisiana (2005)
12. Shi, Y., Zhang, Y., Shih, D.S.: Cloning and expression analysis of two [beta]-1,3-glucanase genes from strawberry. J. Plant Physiol. **163**, 956–967 (2006)
13. Khan, A.A., Shih, D.S.: Molecular cloning, characterization, and expression analysis of two class II chitinase genes from the strawberry plant. Plant Sci. **166**, 753–762 (2004)
14. Van Loon, L.C., Van Strien, E.A.: The families of pathogenesis-related proteins, their activities, and comparative analysis of PR-1 type proteins. Physiol. Mol. Plant Pathol. **55**, 85–97 (1999)
15. Benhamou, N., Kloepper, J.W., Quadt-Hallmann, A., Tuzun, S.: Induction of defense-related ultrastructural modifications in pea root tissues inoculated with endophytic bacteria. Plant Physiol. **112**, 919–929 (1996)

An Alternative Control of Yellow Rust on Bread Wheat with Essential Oils of *Mentha Pulegium*, *Eugenia Aromatica*, and *Cedrus Atlantica*

Marie Solange Uwineza[1(✉)], Brahim El Yousfi[2], and Abdeslam Lamiri[3]

[1] Laboratory of Applied Chemistry and Environment, FST, Hassan 1st University, Settat, Morocco
marie.sol012@gmail.com
[2] Laboratory of Phytopathology, National Institute for Agronomic Research (INRA-CRRA), Settat, Morocco
[3] Superior School of Technology, Hassan 1st University, Berrechid, Morocco

Abstract. Synthetic fungicides have an important role in agriculture development and evolution. However, in intensive agriculture, their misuse threatens directly natural ecosystem stability, and this stability should be protected by relying on other alternatives such as the use of biofungicides. In this regard, the objective of this work was to test three essential oils for the control of yellow rust (*Puccinia striiformis*) of wheats. The experiment used a susceptible variety in a randomized complete block design with four blocks, and four treatments consisted of three essential oils: clove (*Eugenia aromatica*), pennyroyal (*Mentha pulegium*) and Atlas cedar (Cedrus atlantica) along with a chemical treatment (Spiroxamine, Tebuconazole and Triadimenol), applied at a dose of 0.8 L/ha for a slurry of 200 L/ha. The experiment was repeated twice in space under field conditions. The essential oils (1.25 ml/L) as well as the fungicide were applied at the heading stage with a backpack sprayer having a ramp of two meters with four nozzles spaced 0.5 m. The effectiveness of these treatments was evaluated as grain yield increase and thousand-kernel weight (TKW) in comparison to the untreated check. Only pennyroyal essential oil increased grain yield by 23% without affecting TKW, while the fungicide decreased grain yield by 24% by affecting TKW. These results are relative to the concentrations used and to the number of applications, and they prove that pennyroyal essential oil could be an alternative control measure to yellow rust on bread wheat. Implications of these results were discussed in the document.

Keywords: Yellow rust · *Puccinia striiformis* · Bread wheat · Essential oils · *Mentha pulegium*

M. Ezziyyani (Ed.): AI2SD 2018, AISC 911, pp. 54–61, 2019.
https://doi.org/10.1007/978-3-030-11878-5_6

1 Introduction

Agriculture with its diversity and potentialities is considered a key sector of Moroccan economy and it is characterized by cereals, livestock, olive, fruits and vegetable production. With population increase, food needs are increasing and dependency rate, which varies according to agricultural season, shows an upward trend. Demand growth and negative effects of succession of years of acute drought increase economic vulnerability [1].

Morocco is a mediterranean country with a very marked climatic change that affects directly its agriculture by decreasing its capita grain availability. Although cereal area accounts for more than 75% of utilized agricultural area (UAA), their share of grain imports is less than 50%. The strategy of developing this agricultural sector aims to increase production to achieve food self-sufficiency and create jobs. In this context, cereals constitute the most important strategic production [2].

In addition to loss caused by periods of drought, cereals, especially barley and wheat production is hampered by different plant diseases that include septoria, rust, root rot, powdery mildew, and midge. Depending on climatic conditions, cereal diseases if left unchecked can lead to substantial grain yield losses [3, 4].

Recently yellow rust, caused by a fungus Puccinia striiformis took epidemic proportions on bread wheat, and it is recognized by numerous and very small yellow to orange pustules on leaf blades. The fungus would be preserved in the form of uredospores or mycelium in infected stubble during mild winter of Mediterranean regions [5]. Geographical distribution of yellow rust in this country is widespread. Its importance is marked by an oscillating frequency observed between 42 and 93% in wheat fields affected by this disease [6].

To overcome these diseases, farmers have relied on chemical fungicides that certainly improved yields, but their misuse has led in the long term to the development of pathogens resistance, in addition to environmental pollution, many of these fungicides became ineffective or even obsolete. Most synthetic fungicides directly affect essential functions such as respiration, sterol biosynthesis or cell division. This type of mode of action can lead, on the one hand to a substantial risk to humans and non-target organisms and, on the other hand, the development of resistant fungal strains [7–9]. To overcome these constraints, seeking alternative products to these fungicides is needed and remains a necessity in agroecology systems.

Among those alternatives, medicinal plants along with their essential oils can respond to these needs, and industrial use of theses herbs for essential oils (EO), aromas and flavours can contributes not only to alleviate constraints in economies of developing countries by improving exports as well as job creation, and reduction of poverty in vulnerable communities [10], but also improve agriculture production. Some plant extracts have important antibiotic activities and have a broad spectrum of biological properties that are used in pharmaceutical, health and food applications [11–13]. In Morocco, the awareness of the misuse of pesticides in agriculture and their action on environment has led to the use of plant extracts as an alternative to chemical pesticides [14]. Furthermore, developing new molecules from these medicinal plants can be interesting and the orientation of research towards biofungicides formulation seems to be so promising [15].

In this perspective, our tentative work has dealt in situ with yellow rust disease control on a susceptible wheat variety, using essential oils from *Mentha pulegium*, *Eugenia aromatica* and *Cedrus Atlantica* in comparison with a chemical fungicide.

2 Materials and Methods

This experimental study was conducted at the experimental station of INRA Settat: Sidi El Aidi during the agricultural season (2016–2017). In this experiment the used essential oils were extracted from cloves, *Mentha pulegium* leaves and the bark of *Cedrus atlantica* at the Laboratory of Applied Chemistry and Environment of the Faculty of Sciences and Technology of Settat, Morocco. Theses EO were obtained by a steam distillation method using a Clevenger type apparatus and then stored in the dark in opaque glass vials at 4 °C. The EO yield of studied plants was calculated according to (1):

$$EOY = \frac{M'}{M} * 100 \quad (1)$$

Where Y, M' and M represent essential oil yield, essential oil weight and used plant materials weight respectively. Essential oils' chemical composition analysis was performed by CNRST-at Rabat by gas chromatography coupled to mass spectrometer.

A susceptible bread wheat variety Arrihane was used to induce this disease, and experimental plots were mechanically seeded with a Winter Steiger seeder. Yellow rust disease was naturally induced by favorable weather conditions at the end of the growing season [16]. In fact, the disease is sensitive to temperature: high humidities and temperatures between 5 and 15 °C are advantageous for *Puccinia striiformis* infection [5].

The experimental lay out was a randomized complete block design with four blocks. Experimental plots were of 10 m^2 (2 m × 5 m), and seeding was done in 6 rows of 30 cm apart at seeding rate of 120 kg/ha. Before planting, DAP (1 q/ha) was applied to all plots of the experiment which were conducted under supplement irrigation. This supplementary irrigation along with a susceptible bead wheat variety (Arrihane) and climatic conditions of the experimental site were combined to induce natural development of yellow rust. The experiment was repeated twice, in space, at the same experimental station. Two successive years are not necessarily alike concerning the development of the disease, so the two tests were made in the same year [17].

At tillering stage, control of monocotyledonous and some broadleaf weeds was achieved by an application of 0.5 L/ha of a systemic herbicide (45 g/L Pyroxsulam composition, an active substance of the chemical family of triazolopyrimidines). One week later, we applied 2, 4-D (2, 4-dichlorophenoxyacetic acid) at a dose of 1L/ha to achieve a complete herbicide control. Ammonitrate (33%) was applied as top dressing at a rate of two application (1 q/ha each) one week after the second weed-control passage and a second one at heading stage. The EO and the chemical treatments were applied once [18] at late heading stage [19].

At the heading stage, a backpack sprayer with a two-meter boom that has four nozzles spaced of 0.5 m, was used to apply these four treatments, The three essential oils pennyroyal, cloves and cedar of atlas were applied at 1.25 ml/L (water), while the chemical fungicide (Spiroxamine, Tebuconazole and Triadimenol) was applied 0.8 L/ha as a slurry of 200 L/ha. At maturity, four canter rows of each experimental plot were mechanically harvested with a plot combiner. Yellow rust control was evaluated as grain yield and thousand kernel weight differences between treatments [20–22].

Data were analyzed using SPSS software ANOVA (analysis of variance) procedure, and paired comparison of the treatments to the check was performed at 5% probability. Yield gain was calculated according to (2), where TY and NTY refer to average yield of treated plots and average yield of untreated control plots, respectively.

$$\frac{TY - NTY}{NTY} * 100 \tag{2}$$

3 Results and Discussions

We noticed that grain yield and TKW were affected by the disease as detected in [23–25]. Compared to the non treated check, mean grain yield of treated plots with essential oils were greater, while grain yield of the chemical treatment was lower. Essential oil of pennyroyal increased yield by 23% when compared to the check and this result was in accordance with results obtained in [18] with a concentration 2 ml/L of the essential oils of eight aromatic plants. However, in our experiment the fungicide treatment decreased grain yield by 24%, and application of these 3 essential oils did not affect significantly thousand-kernel weight, while the fungicide treatment did (Table 1).

Table 1. Effects of essential oils from *Mentha pulegium*, *Eugenia aromatica*, *Cedrus atlantica* and chemical fungicide on the plot yield

Treatments	Plot yield (Kg)	Performance gain (%)	Thousand kernel weight (g)
Mentha pulegium	2.33a	22.60	35.40^{b}
Eugenia aromatica	1.92^{b}	1.05	35.62^{b}
Cedrus atlantica	1.84^{c}	−3.15	34.28^{b}
Fungicide	1.45^{d}	−23.70	32.50^{a}
Check	1.90^{b}	–	34.79^{b}

Note: Values within the same column and followed by the same letters are not significantly different at 5% probability.

Despite the advances in formulation and application, fungicides continue to represent an option with serious drawbacks. However, in this regard, our fungicide application aiming to control wheat stripe rust under semi-arid region and water stress had negative impact on yield, even though it did reduce disease severity. In this situation and in addition to the concerns for environmental damage, use of essential oils may take advantage over chemical control of plant disease.

So far, a major alternative disease control method for this disease is the development of wheat cultivars that are resistant to current populations of the pathogen [26]. However and worldwide, the current status of this wheat constraint would suggest that resistance breeding alone fails to contain stripe rust epidemics due to a macrocyclic lifecycle of Puccinia striiformis [27], and application of essential oils, when coupled with resistance, may alleviate this constraint by hampering emergence of new races as those lately noticed in Europe [28, 29].

Application of essential oils to control this disease may also be a component of an integrated control measure with improved control strategies and practices [30, 31] and used like effective bio-fungicides that should be developed for use in fields under organic production or in areas where restrictions to chemicals is enhanced for environment protection.

Studies dealing with vivo antifungal activities of many plant extracts against plant pathogenic fungi can give new insights to disease control [32]. In this regard, the antimicrobial activity of the essential oil of Mentha pulegium could be attributed to its major constituents, which are monocyclic oxygenated monoterpenes. The mechanism of action of terpenes is not fully understood, but it is likely that these lipophilic compounds cause a loss of integrity of the membrane of fungi [33]. Thus, antifungal properties of essential oils used in this study could well be attributed to their major components [34–36] but minor components can also be involved through a synergistic effect [37, 38]. This essential oil may also increased both incubation period and latent period (LP) of the disease. It may decreased the number of pustules/cm2 as in a similar study under greenhouse and field conditions using some essential oils, i.e. chamomile, thyme, cumin, basil, eucalyptus and garlic oils to control wheat rust disease at seedling and adult stage of two susceptible wheat cultivars such as Morocco and Sids-1 [39].

4 Conclusion

The results developed herein are related to the concentrations used, and essential oil of *Mentha pulegium* gave a better result by increasing grain yield by 23% and could thus constitute an alternative biofungicide to the chemical which unfortunately induce a grain yield loss of 24% by lowering thousand kernel weights. This decrease might came from susceptibility of the plants experiencing a heat stress during grain filling, the thing that was not noted in plots treated with EO. The essential oils of *Eugenia aromatica* and *Cedrus atlantica* did not produce expected results. We think that higher concentrations of these EO and more than one application, may lead to better results in controlling yellow rust. In addition, to bring this disease under control, a coordinated strategy for rust control, that includes essential oils application, needs to be implemented.

Acknowledgements. ARGB thanks technicians of the Sidi El Aidi experimental station and the technical assistance of Fatna Ajouad in the Cereal Pathology Laboratory of INRA Settat.

References

1. Ministère de l'Agriculture et de la Pêche Maritime: Les filières phares du secteur de l'Agriculture, L'agriculture Marocaine en chiffre, pp. 12–30 (2012)
2. Aït Hamza, M.: Les céréales dans le Maroc du Centre-ouest. Méditérranée **88**, 27–32 (1998)
3. Rieuf, P., Teasca, G.: Etudes sur les helmenthosporium des céréales du Maroc. Al Awamia **46**, 29–58 (1973)
4. El Yousfi, B.: Guide du diagnostic des principales maladies des céréales d'automne au Maroc. INRA-Centre Régional de la Recherche Agronomique, Settat (2015)
5. Nasraoui, B.: Principales maladies fongiques des céréales et des légumineuses en Tunisie. Centre de Publication Universitaire, Kef (2008)
6. Arifi, A.: Évolution et importance des maladies fongiques du blé au Maroc. Al Awamia **90**, 19–30 (1995)
7. Kessmann, H., Staub, T., Hofmann, C., Maetzke, T., Herzog, J., Ward, E., Uknes, S., Ryals, J.: Induction of systemic acquired disease resistance in plants by chemicals. Ann. Rev. Phytopathol. **32**, 439–459 (1994)
8. Koffi, A.G., Komlan, B., Kouassi, A., Mireille, P., Messanvi, G., Philippe, B., Koffi, A.: Activité antifongique des huiles essentielles de *Ocimum basilicum* L. (Lamiaceae) et *Cymbopogon schoenanthus* (L.) Spreng. (Poaceae) sur des micromycètes influençant la germination du maïs et du niébé. Acta Botanica Gallica **153**(1), 115–124 (2013)
9. Tasei, J.N.: Impact des pesticides sur les abeilles et autres pollinisateurs. Courrier de l'environnement de l'INRA, pp. 9–18 (1996)
10. Okigbo, R.N., Anuagasi, C.L., Amadi, J.E.: Advances in selected medicinal and aromatic plants indigenous to Africa. J. Med. Plants Res. **III**(2), 086–095 (2009)
11. Yang, V.W., Clausen, C.A.: Inhibitory effect of essential oils on decay fungi and mold growth on good. Am. Wood Prot. Assoc. **103**, 62–70 (2007)
12. El Ajjouri, M., Satrani, B., Ghanmi, M., Aafi, A., Farah, A., Rahouti, M., Amarti, F., Aberchane, M.: Activité antifongique des huiles essentielles de *Thymus bleicherianus* Pomel et *Thymus capitatus* (L.) Hoffm. & Link contre les champignons de pourriture du bois d'œuvre. Biotech. Agron. Soc. Environ. **12**(4), 345–351 (2008)
13. Aissaoui, B.A., Zantar, S., Toukour, L., El Amrani, A.: Etude de la composition chimique de l'huile essentielle de *Rosmarinus officinalis* et évaluation de son efficacité sur l'acarien ravageur *Tetranychus urticae*. Association Marocaine de Protection des Plantes, pp. 55–64 (2016)
14. El Guilii, M., Achbani, E., Fahad, K., Jijakli, H.: Biopesticides: Alternatives à la lutte chimique? In: Symposium international "Agriculture durable en région Méditéranéenne", Rabat (2009)
15. Amri, I., Hamrouni, L., Hanana, M., Gargouri, S., Jamoussi, B.: Propriétés antifongiques des huiles essentielles de *Biota orientalis* L. Phytothérapie **12**, 170–174 (2014)
16. El Jarroudi, M.: Evaluation des paramètres épidémiologiques des principales maladies cryptogamiques affectant les feuilles du blé d'hiver au Grand-Duche de Luxembourg: calibration et validation d'un modèle de prévision, Thèse de Doctorat, Université de Liège (2005)
17. Duvivier, M., Matieu, O., Heens, B., Meza, R., Monfort, B., Legrève, A., Seutin, B., Bodson, B., Deproft, M.: Lutte intégrée contre les maladies. Livre Blanc "Céréale", pp. 1–59 (2013)
18. Shabana, Y.M., Abdalla, M.E., Shahin, A.A., El-Sawy, M.M., Draz, I.S., Youssif, A.W.: Efficacy of plant extracts in controlling wheat leaf rust disease caused by *Puccinia triticina*. Egypt. J. Basic Appl. Sci. **4**, 67–73 (2017)

19. Alim'agri: Reconnaître au champ la Rouille jaune (*Puccinia striiformis*). Perspectives agricoles, 1398, 41–42 (2013)
20. Charles, R., Cholley, E., Frei, P.: Assolement, travail du sol, variété et protection fongicide en production céréalière. Recherche Agronomique Suisse **5**(2), 212–219 (2011)
21. Vergara-Diaz, O., Kefauver, S.C., Elazab, A., Nieto-Taladriz, M.T., Araus, J.L.: Grain yield losses in yellow-rusted durum wheat estimated using digital and conventional pameters under field conditions. Crop J. **3**, 200–211 (2015)
22. Hailu, D., Fininsa, C.: Relationship between stripe rust (*Puccinia striiformis*) and common wheat (*Triticum aestivum*) yield loss in the highlands of Bale, southeastern, Ethiopia. Arch. Phytopathol. Plant Prot. **42**(6), 508–523 (2009)
23. Afzal, S.N., Haque, M.I., Ahmedani, M.S., Bashir, S., Rattu, A.U.R.: Assessment of yield loss caused by *Puccinia striiformis* triggerings stripe rust in the most common wheat varieties. Pak. J. Bot. **39**(6), 2127–2134 (2007)
24. Sajid, A., Syed, J.A.S., Hidayatur, R., Saquib, M.S., Ibrahim, M., Sajjad, M.: Variability in wheat yield under yellow rust pressure in Pakistan. Turk. J. Agric. For. **33**, 537–546 (2009)
25. Wendale, L., Ayalew, H., Woldeab, G., Mulugeta, G.: Yellow rust (*Puccinia striiformis*) epidemics and yield loss assessment on wheat and triticale crops in Amhara region, Ethiopia. Afr. J. Crop Sci. **4**(2), 280–285 (2016)
26. Wellings, C.R., Boyd, L.A., Chen, X.M.: Resistance to stripe rust in wheat: Pathogen biology driving resistance breeding. In: Disease Resistance in Wheat, Nosworthy Way Wallingford Oxfordshire, CABI, p. 335 (2012)
27. Wellings, C.R.: Global status of stripe rust: a review of historical and current threats. Euphytica **179**, 129–141 (2011)
28. Cheyron, P., Maufras, J.Y., Audigeos, D., Vallavieille-Pope, C., Leconte, M.: Rouille jaune: une race s'attaque aux céréales. Perspectives agricoles **410**, 1–18 (2014)
29. Wan, A., Wang, X., Kang, Z., Chen, X.X.: Variability of the stripe rust pathogen. In: Stripe rust, p. 723. Springer Science + Business Media B.V., Dordrecht (2017)
30. McIntosh, R., Wellings, C., Park, R.F.: Wheat Rusts: An Atlas of Resistance Genes, p. 213. CSIRO Publications, Clayton (1995)
31. Chen, X., Kang, Z.: Stripe rust research and control: conclusions and perspectives. In: Stripe Rust, p. 723. Springer Science + Business Media B.V, Dordrecht (2017)
32. Choi, G.J., Jang, K.S., Kim, J.S., Lee, S.W., Cho, J.Y., Cho, K.Y., Kim, J.C.: In vivo fungal activity of 57 plant extracts against six plant pathogenic fungi. Plant Pathol. J. **20**(3), 184–191 (2004)
33. Hmiri, S., Amrani, N., Rahouti, M.: Détermination in vitro de l'activité antifongique des vapeurs d'eugénol et d'huiles essentielles de *Mentha pulegium* L. et de *Tanacetum annuum* L. vis-à-vis de trois champignons responsables de la pourriture des pommes en post-récolte. Acta Botanica Gallica **158**(4), 609–616 (2011)
34. Soltani, N., Kellouche, A.: Activite´ biologique des poudres de cinq plantes et de l'huile essentielle d'une d'entre elles sur *Callosobruchus maculatus* (F.). Int. J. Trop. Insect Sci. **24** (2), 184–191 (2004)
35. Bourkhiss, M., Hnach, M., Bourkhiss, B., Ouhssine, M., Chaouch, A.: Composition chimique et propriétés antimicrobiennes de l'huile essentielle extraite des feuilles de *Tetraclinis articulata* (Vahl) du Maroc. Afrique Sci. **03**(2), 232–242 (2007)
36. Amarti, F., Satrani, B., Ghanmi, M., Farah, A., Aafi, A., Aarab, L., El Ajjouri, M., Chaouch, A.: Composition chimique et activité antimicrobienne des huiles essentielles de *Thymus algeriensis* Boiss. & Reut. et *Thymus ciliatus* (Desf.) Benth. du Maroc. Biotechnologie, Agronomie, Société et Environnement **14**(1), 141–148 (2010)

37. Kordali, S., Cakir, A., Oze, H., Cakmakci, R., Kesdek, M., Mete, E.: Antifungal, phytotoxic and insecticidal properties of essential oil isolated from turkish *Origanum acutidens* and its three components, carvacrol, thymol and p-cymene. Biores. Technol. **99**, 8788–8795 (2008)
38. Bouzouita, N., Kachouri, F., Ben Halima, M., Chaabouni, M.M.: Composition chimique et activités antioxydante, antimicrobienne et insecticide de l'huile essentielles de *Juniperus phœnicea*. Journal de la Société Chimique de Tunisie **10**, 119–125 (2008)
39. Tohamey, S., El-Sharkawy, H.H.A.: Effect of some plant essential oils against wheat leaf rust caused by *Puccinia triticina* f. Sp tritici. Egypti. J. Bio. Pest Control **24**(1), 211–216 (2014)

Study of Growth and Production of *Botrytis Cinerea* Conidia of Some Morrocan Isolates in Different Nutrients Media

Ahlem Hamdache[1(✉)], Mohammed Ezziyyani[2], and Ahmed Lamarti[1]

[1] Faculty of Sciences of Tetouan, Department of Biology, Abdelmalek Essaâdi University, Avenue de Sebta, Mhannech II, 93002 Tétouan, Morocco
hamdach_ahlem@yahoo.fr

[2] Polydisciplinary Faculty of Larache, Department of Life Sciences, Abdelmalek Essaâdi University, 745 Poste Principale, 92004 Larache, Morocco

Abstract. *Botrytis* species include some serious fungal plant pathogens, which are implicated in many diseases affecting flowers, fruits, cereals, legumes, and other vegetables. In particular, *Botrytis cinerea* attacks economically important crops such as carrots, grapes, lettuce, strawberries, and tobacco, producing various leaf spot diseases and grey mould. In this work, we have studied, *in vitro*, the growth and the sporulation of some isolates of *Botrytis cinerea* in different nutrient media. Our aim was to select the isolate the most pathogen to use it in biological control tests. For this purpose, ten isolates of *Botrytis cinerea* were isolated, purified and identified in the laboratory. Seven isolates were isolated from samples of strawberry fruit harvested from fields; four of them, are originated from strawberry fields of Zlaoula and Laâwamra (Larache) and three are originated from fields of Moulay Bouselham (Algharb). Others are isolated from postharvest strawberry. The growth and sporulation of all of the isolates are studied and compared in different nutrient media PDA, MEA, Czapek and organic medium of strawberry. Among all of the isolates tested, *Botrytis cinerea* Bt7 originated from fields of Zlaoula was the most important isolate with maximum growth in all of nutrients media, and maximum sporulation ($74{,}7.10^5$ sp/ml). *B. cinerea* was inoculated artificially in leaf and fruit stawberry and it causes the visible disease symptoms of grey mould.

Keywords: *Botrytis cinerea* · Nutrient media · Mycelial growth · Sporulation · Pathogenicity

1 Introduction

Fungi of the genus Botrytis Persoon are important pathogens of many agronomically important crops, such as strawberry, grapevine, tomato, bulb flowers, and ornamental crops [1]. *Botrytis cinerea* is a pathogenic ascomycete responsible for grey mould on a diversity of plant tissue types across hundreds of dicotyledonous plant species [2]. *Botrytis* diseases appear primarily as blossom blights and fruit rots but also as leaf spots and bulb rots in the field and in stored products. Botrytis species are necrotrophs,

M. Ezziyyani (Ed.): AI2SD 2018, AISC 911, pp. 62–68, 2019.
https://doi.org/10.1007/978-3-030-11878-5_7

inducing host-cell death resulting in progressive decay of infected plant tissue. *B. cinerea* is a capable saprotroph and necrotroph, with different genetic types often showing a trade-off between saprotrophic and necrotrophic capabilities [3]. The pathogen produces abundantly sporulating gray mycelium on infected tissue. Macro-conidia (mitotically produced spores) can be transported by wind over long distances. Botrytis overwinters in the soil as mycelium in decaying plant debris and as sclerotia, melanized mycelial survival structures. Sporulation is a process that involves the development of sexual or asexual spores and associated structures. It is considered that this mechanism of reproduction is controlled by genetic, hormonal, nutritional and environmental factors. However, virulence levels of *B. cinerea* strains are not necessarily a fixed feature. For example, virulence has been observed to diminish during protracted *in vitro* culture [4]. Degenerated cultures have been reported in a wide range of pathogenic fungi [5], although very little is known about why cultures lose virulence. The survival of fungal species depends in part on its ability to produce large quantities of viable conidia under different environmental conditions and even physical and nutritional requirements are more stringent than those required for mycelial growth [6]. For the development of sporulation there are two major nutrient requirements that are a source of carbon and nitrogen source which is required for synthesizing amino acids, proteins, and nucleic acids necessary for the construction of protoplasm [7]. Glucose is the carbon source most widely used by the fungi followed by fructose, mannose, galactose, sucrose, lactose, maltose and polysaccharides such as starch, cellulose, pectin [8], chitin [9], trehalose, sorbitol and mannitol [10]. A commercial level has developed a wide variety of agar formulations to induce sporulation and the growth of fungi; among the culture media commonly used we can found the Malt extract agar (MEA), Nutrient agar (NA), Potato dextrose agar (PDA), to name a few [11]. Most of these present a media composition based on source of carbon and one nitrogen which generally vary between them and in the proportions in which they are presented as well as in its origin which may be organic or inorganic. The initial stages of isolation and morphological characterization of new isolates is still of vital importance for the use of plate assays because the understanding of essential aspects of fungal growth can be very useful in projects for finding a good methods of biological control pathogenic fungi. In order to determine the ability of *Botrytis cinerea* to grow and sporulate in four different culture media, in the present study we evaluated ten native strains of *Botrytis cinerea* fungus isolated from strawberries cultures of Morocco. The growth and sporulation of the ten isolates were compared for selecting the isolate the most aggressive for use it in biological control tests; the isolate selected must having an important growth and sporulation.

2 Materials and Methods

2.1 Samples

Isolates of *Botrytis cinerea* have been isolated from samples of strawberry fruits with symptoms of gray mold. Several strains are taken from different strawberry plants grown in different fields in the area of Zlaoula, Loukkos (Larache) and Moulay

Bousselham (El Gharb). Other isolates are isolated from post-harvested strawberries. Isolates of *Botrytis spp.* were isolated, purified and identified. They were been isolated from the surface of the fruit. It has been grown on the PDA (Potato DextroseAgar, Biokar Diagnostics) culture medium. After several subcultures, we purified ten isolates of *Botrytis cinerea*. The isolate was maintained at 4 °C in the PDA medium. The isolates of *Botrytis cinerea* species used in this study were listed in Table 1.

Table 1. Origin of *Botrytis cinerea* isolates.

Pre-harvested isolates Loukkos (Larache)	Pre-harvested isolates Moulay Bouselham (El Gharb)	Post harvested isolates
Bt3	A1	Bt1
Bt5	M2	Bt4
Bt7	M3	Bt8
Bt10		

2.2 Fungal Identification

The taxonomic identification of species was followed, from the isolation media, according to the macro and microscopic characteristics of the colonies using adequate identification keys [12, 13].

2.3 Media Preparation

Some of the common synthetic, semi synthetic and natural media, in solid form, were used to culture the fungus. Purified culture of the fungus was inoculated into the five different solid agar media, namely, potato dextrose agar (PDA), malt extract agar (MEA), Czapek and natural medium at base of strawberry.

- Potato Dextrose Agar (PDA)

The composition of the PDA medium was 200 g of Potato, 20 g of Glucose, 16 g of Agar in 1000 ml of Distilled water.

- Malt Extract Agar (MEA)

The composition of the MEA medium was 20 g of Malt Extract 20 g Glucose, 1 g Peptone, 15 g Agar in 1000 ml of Distilled water.

- Czapek

The composition of the Czapek medium was 2 g of $NaNO_3$, 1 g of K_2HPO_4, 0,5 g of $MgSO_4x7H_2O$, 0,5 g of KCl, 0,01 g of $FeSO_4$ $x7H_2O$, 30 g of Saccharose, 16 g of Agar in 1000 ml of Distilled water.

- Strawberry

Strawberry and distilled water.

2.4 Mycelial Growth

For each medium, three plates were inoculated by mycelial discs of each of *Botrytis cinerea* isolates. The plates are incubated at 25 °C. Growth was recorded after 10 days of incubation by measuring the diameter of the mycelial colony.

2.5 Sporulation

Sporulation was assessed using a hemocytometer. Conidia were gently removed with a bacteriological loop suspended in sterile distilled water containing 0.01% Tween-20, and filtered through sterile cheesecloth to remove remaining mycelia. The concentration was determined with a hemocytometer.

3 Results and Discussions

The mycelia growth of *Botrytis cinerea* was favorable in PDA and strawberry medium for all of the isolates, whereas was poor in Czapek. The Bt7, M2 and M3 isolates have a maximum mycelial growth on all of the culture media “Fig. 1”.

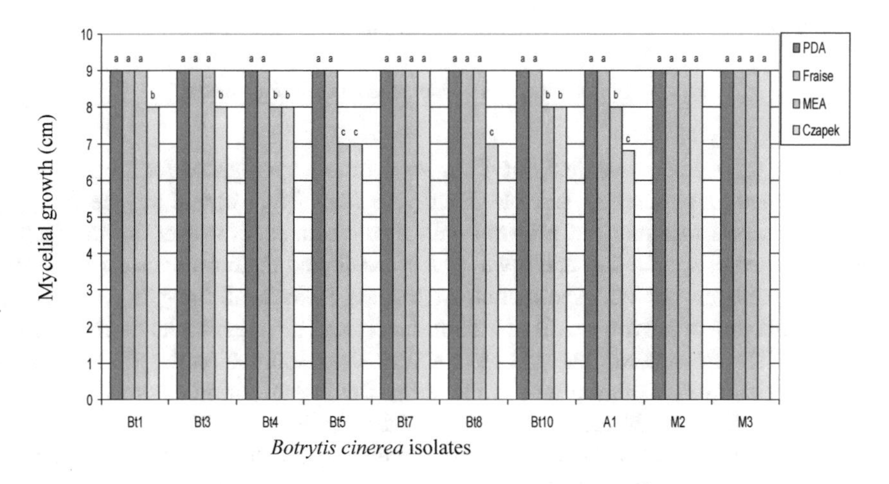

Fig. 1. Mycelial growth of B. cinerea isolates in four media.

After 10 days of incubation (at 25 °C), all of *Botrytis cinerea* isolates, each having a different origin of isolation, showed maximum growth (9 cm diameter) on the two media PDA and strawberry. On MEA and Czapek medium, growth of the isolates was variable from one strain to another, ranging from 6.8 (isolate A1) to 9 cm (Bt7, M2 and M3) on Czapek medium, and 7 (Bt5) to 9 cm (Bt1, Bt3, Bt7, Bt8, M2 and M3) on MEA medium (Fig. 1).

There is a significant difference of sporulation among isolates. The majority has a sporulation equal or lower to 20×10^5 spores/ml (except the A1 isolate with 28.66×10^5 spores/ml and the Bt7 isolate with 74.7×10^5 spores/ml on the PDA medium). The average production of conidia obtained in PDA medium ranged from 3×10^5 (for the isolate Bt10) to 74.7×10^5 (for the isolate Bt7) spores/ml. Some isolates showed a weak sporulation like M3 (4.4×10^5 spores/ml) to very low as Bt10. (3×10^5 spores/ml). The Bt7 isolate showed a sporulation higher than other isolates (74.7×10^5 spores/ml on the PDA medium) and 30.25×10^5 spores/ml on Strawberry medium (Fig. 2).

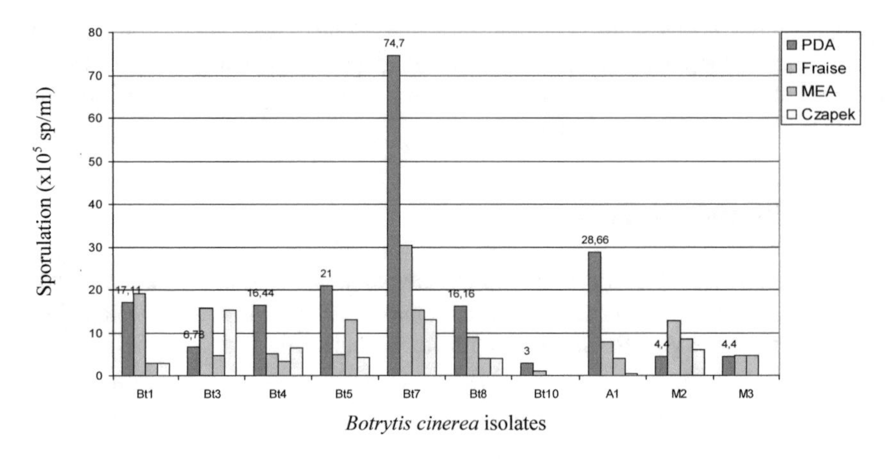

Fig. 2. Sporulation of B. cinerea isolates in four media.

Among the mediums studied, the PDA seems most appropriate for all isolates of *Botrytis cinerea* and the most suitable. Finally, isolates Bt7, M2 and M3 are the only ones to present a maximum of growth on all media studied. However the Bt7 isolate presents a greater sporulation and it was being considered the most aggressive isolate. Comparing the growth and sporulation of the ten isolates of *Botrytis cinerea*, we observed maximum growth on all of culture media and a well sporulation marked for the isolate Bt7. Isolates M2 and M3 have a very weak sporulation despite their maximum mycelial growth, reason why we kept the Bt7 isolate for other studies. Several authors have found a difference between isolates of the same species; The natural isolates of *B. cinerea* do not present the same level of aggression on the same host plant [14–16]. *B.cinerea*, requires exogenous nutrients to produce aggressive lesions as has been shown for other isolates [17, 18]. Nutrients are necessary for the germination of spores, mycelium development, and training of the appressoria [19]. Ten isolates of *Botrytis cinerea* were grown on four culture media. The PDA medium considered to be better than others for mycelial growth and sporulation of *Botrytis cinerea*. According to the bibliography, the MEA has been used by some researchers [20], but the PDA is the most used nutrient medium for cultivation of *Botrytis cinerea*. [21–23].

4 Conclusion

The objective of this work was to determine, *in vitro*, the most virulent isolate of *Botrytis cinerea* among ten moroccan isolates from different origin, in different medium, comparing their mycelial growth and sporulation in order to use it in biological control of grey mould. The results strongly support the idea that the use of Bt7 isolate of *B. cinerea*, as a most virulent agent, in other works of biological control will give a best results in this purpose than a less virulent isolate. The PDA medium was the most favorable medium for mycelial growth and conidia production. For this reason it can be selected for establishing the *in vitro* interactions between the pathogen *B. cinerea* and the antagonist in biological control tests.

References

1. Jarvis, W.R.: Botryotinia and *Botrytis* species; taxonomy, physiology and pathogenicity. Monograph No. 15, Canadian Department of Agriculture, Ottawa, p. 195 (1977)
2. Fournier, E., Gladieux, P., Giraud, T.: The "Dr Jekyll and Mr Hyde fungus": noble rot versus gray mold symptoms of *Botrytis cinerea* on grapes. Evol. Appl. **6**, 960–969 (2013)
3. Martinez, F., Dubos, B., Fermaud, M.: The role of saprotrophy and virulence in the population dynamics of botrytis cinerea in vineyards. Phytopathology **95**, 692–700 (2005)
4. Pathirana, R., Cheah, L.H., Carimi, F., Carra, A.: Low temperature stored in cryobank® maintains pathogenicity in grapevine. Cryoletters **30**, 84 (2009)
5. Butt, T.C., Wang, S.F., Hall, R.: Degeneration of entomogenous fungi. In: Eilenberg, J., Hokkanen, H. (eds.) An Ecological and Societal Approach to Biological Control, pp. 213–226. Springer, Dordrecht (2006)
6. Moore, E.: Fundamentals of the Fungi. Prentice Hall, New Jersey, p. 574 (1996)
7. Pérez, L., Ramírez, C.: Effect of the variables, conditions of fermentation and the substrate in the production of Trichoderma harzianum. Ph.D. dissertation, Pontificia Universidad Javeriana, Bogotá, Colombia, p. 153 (2000)
8. Griffin, D.H.: Fungal Physiology, pp. 260–279. John Wiley & Sons, Inc., New Jersey (1996)
9. Hegedus, D., Bidochka, M., Khachatourians, G.: *Beauveria bassiana* submerged conidia production in a defined medium containing chitin, two hexosamines or glucose. Appl. Microbiol. Biotechnol. **33**, 641–647 (1990)
10. Bidochka, M., Low, N., Khachatourians, G.: Carbohydrate storage in the entomopathogenic fungus *Beauveria bassiana*. Appl. Environ. Microbiol. **56**, 3186–3190 (1990)
11. Kamp, A.M., Bidochka, M.J.: Conidium production by insect pathogenic fungi on commercially available agars. Lett. Appl. Microbiol. **35**, 74–77 (2002)
12. Samson, R.A., Hoekstra, E.S., C.A.N.V. Oorschot: Introduction to food-borne fungi. 2nd ed. Centraalbureau Voor Schimmelcultures, BAARN. Institute of the Royal Netherlands, Academy of Arts and Sciences, p. 248 (1984)
13. Botton, B., Breton, A., Fevre, M., Gauthier, S., Guy, P., Larpent, J.P., Reymond, P., Sanglier, J.J., Vayssier, Y., Veau, P.: Moisissures utiles et nuisibles: importance industrielle. 2e éd. rev. et compl. Paris, Milan, Barcelone, Masson, p. 512 (1990)
14. Tiedemann, A.V.: Evidence for a primary role of active oxygen species in induction of host cell death during infection of bean leaves with *Botrytis cinerea*. Physiol. Mol. Plant Pathol. **50**(3), 151–166 (1997)

15. Decognet, V., Bardin, M., Trottin-Caudal, Y., Nicot, P.C.: Rapid change in the genetic diversity of Botrytis cinerea populations after the introduction of strains in a tomato glasshouse. Phytopathology **99**, 185–193 (2009)
16. Mirzaei, S., Mohammadi-Goltapeh, E., Shams-Bakhsh, M., Safaie, N., Chaichi, M.: Genetic and phenotypic diversity among Botrytis cinerea isolates in Iran. J. Phytopathol. **157**, 474–482 (2009)
17. Kosuge, T., Hewitt, W.B.: Exudates of grape berries and their effect on germination of conidia of Botrytis cinerea. Phytopathology **54**, 167–172 (1964)
18. Yoder, O.C., Whalen, M.L.: Factors affecting postharvest infection of stored cabbage tissue by *Botrytis cinerea*. Can. J. Bot. **53**, 691–699 (1975)
19. Li, G.Q., Huang, H.C., Acharya, S.N., Erickson, R.S.: Biological control of blossom blight of alfalfa caused by *Botrytis cinerea* under environmentally controlled and field conditions. Plant Dis. **88**, 1246–1251 (2004)
20. Edwards, S.G., Seddon, B.: Mode of antagonism of Brevibacillus brevis against *Botrytis cinerea* in vitro. J. Appl. Microbiol. **91**, 652–659 (2001)
21. Hjeljord, L.G., Stensvand, A., Tronsmo, A.: Antagonism of nutrient-activated conidia of Trichoderma harzianum (atroviride) P1 against *Botrytis cinerea*. Phytopathology **91**(12), 1172–1180 (2001)
22. Guetsky, R., Shtienberg, D., Elad, Y., Dinoor, A.: Combining biocontrol agents to reduce the variability of biological control. Phytopathology **91**(7), 621–627 (2001)
23. Buck, J.W.: *In vitro* antagonism of *Botrytis cinerea* by phylloplane yeasts. Can. J. Bot. **80**(8), 885–891 (2002)

Data Mining for Predicting the Quality of Crops Yield Based on Climate Data Analytics

Maroi Tsouli Fathi(✉), Mostafa Ezziyyani, and Soumaya El Mamoune

Faculty of Science and Technology/Computer Science, Tangier, Morocco
maroi.tsouli@gmail.com, ezziyyani@gmail.com, soumayamgi@gmail.com

Abstract. This study assesses and predicts the impact of climate change on the harvest of agricultural crops in Morocco using the data mining approach. Several econometric models have been tested based on primary data. These models made it possible to establish part of the relationship between agricultural income and climatic variables (temperature and precipitation) and, on the other hand, to analyze the sensitivity of agricultural incomes to these climatic variables. The field of agriculture is extremely sensitive to the change of the climate, the variations intra and inter-seasonal cause the increase in the temperatures and the variations on the modes of precipitation which decreases the seasonal crop yields and increases the probability of bad short-term harvests and a reduction of the long-term production. However, this relation between climate change and agriculture are not yet foreseeable for the future, it will be thus interesting to make a predictive study which will allow the climatic analysis of data followed by an Agro climatic study of data to establish the connection between climate change and agricultural production and suggested afterward plans of adaptation to this change. In this study, we will carry out a comparative study, between the various methodology and tools of analysis of data of data mining to choose the algorithms that will adapt the best for our predictive analysis which will allow us to determine the threat of the impact of the climate change on the production of certain agricultural crops in morocco.

Keywords: Morocco · Agriculture · Climate change · Data mining algorithm · Data analysis

1 Introduction

The agricultural sector accounts for 15 to 20% of the annual GDP in Morocco [16]. In the early 2000s, Morocco began to study climate change and conducted several prospective studies describing the expected impacts of climate change on climate change agriculture [13]. A lot of strategy and project have seen the day such as the green Morocco plan or even more the mosaic project.

Based on climate studies and statistics, the country's agriculture appears vulnerable to the effects of climate change.

M. Ezziyyani (Ed.): AI2SD 2018, AISC 911, pp. 69–79, 2019.
https://doi.org/10.1007/978-3-030-11878-5_8

Morocco is an industrial country but relies heavily on the agricultural sector more than 12.25% of the country's surface is farmed for agriculture.

Among the main Moroccan crops, cereals (wheat, barley), citrus fruits (oranges, clementines), olives, fruit rosaceae (almonds, apples, apricots…), sugar beets, food legumes, crops market gardening including potatoes and tomatoes, but this natural environment is threatened by several climatic handicaps such as low rainfall. Weather conditions can cause significant variations in agricultural production from one year to the next, leading the state to become involved in climate risk management through insurance schemes. Moroccan agriculture is particularly sensitive to climate change. A quarter of the total area of Morocco is already threatened by desertification.

2 The Problem Domain

Climate change will generally have negative impacts on agriculture and threaten food security in Morocco, variable effects on the performance of irrigated and non-irrigated crops across regions, declines in production of certain agricultural crops rising prices of key agricultural products: rice, wheat, corn, soybeans, olives…

The aim of our work and the use of predictive analytics associated with detailed modeling of crop growth under climate change, based on two scenarios to simulate the climate and performance. The results of this analysis suggest solutions for the adaptation of certain agricultural crops.

To know:

- Climate Influence on seasonal crops
- Estimated crop yield
- Estimated damage caused by climate change
- Proposals adaptation solutions to the new climate.

3 Methodologies

The methodology to adapt for our predictive system will consist of two parts:

- The consists of the analysis of climate data (historical and forecast) to evaluate the new climate, followed by the follow-up and studies of observed trends to construct an assessment of the impact of climate change or change of weather
- secondly, the second part of the simulator will consist of evaluating the impact of CC on agricultural yields and the evaluation of performance-based meteorological parameters.

3.1 Issues Related to Data

The climate database is complex and contains several parameters that are not related to each other these data do not follow any precise logic or rules so that I can apply on datamining algorithms [9].

Then it will be preferable to generate from this data a new database of binary type more optional.

The olive tree and says a Mediterranean type culture and it does not support the humidity from where we will build a new data base from the monthly forecasts that follows the rules of the Mediterranean climate [7].

3.2 Description Features

See Table 1.

Table 1. Description climate features

Computed	✓ Cooling Degree Days (CLDD) ✓ Heating degree days (HTDD) ✓ Number days with maximum temperature < 32 F. (DX32) ✓ Number days with maximum temperature > 70 F (21.1C) (DX70) ✓ Number days with maximum temperature > 90 F (32.2C) (DX90) ✓ Number days with minimum temperature less than or equal to 0.0 F(DT00) ✓ Number days with minimum temperature less than or equal to 32.0 F(DT32) ✓ Number of days with greater than or equal to 0.1 inch of precipitation(DP01) ✓ Number of days with greater than or equal to 0.5 inch of precipitation(DP05) ✓ Number of days with greater than or equal to 1.0 inch of precipitation(DP10)
Precipitation	✓ Extreme maximum precipitation for the period. (EMXP) ✓ Precipitation (PRCP)
Air Temperature	✓ Average Temperature. (TAVG) ✓ Cooling Degree Days Season to Date (CDSD) ✓ Extreme maximum temperature for the period. (EMXT) ✓ Extreme minimum temperature for the period. (EMNT) ✓ Heating Degree Days Season to Date (HDSD) ✓ Maximum temperature (TMAX) Minimum temperature (TMIN)

3.3 Rules of the Mediterranean Climate

The climatic classification of Köppen has been developed on the basis of the relationship between climate and vegetation. From this classification it is possible to define the climatic conditions for each region and subsequently establish a relationship between climate and suitable vegetation. This approach was established by Koppen based on multiple seasonal coefficients.

Koppen's classification has been used in several research works concerning the phenomenon of climate change and calculates its evolution, the diagnostics carried out from the classification of Koppen have given very correct results over the years and it is even use now to monitor his evolution has deferent time scales (Fig. 1 and Tables 3, 4).

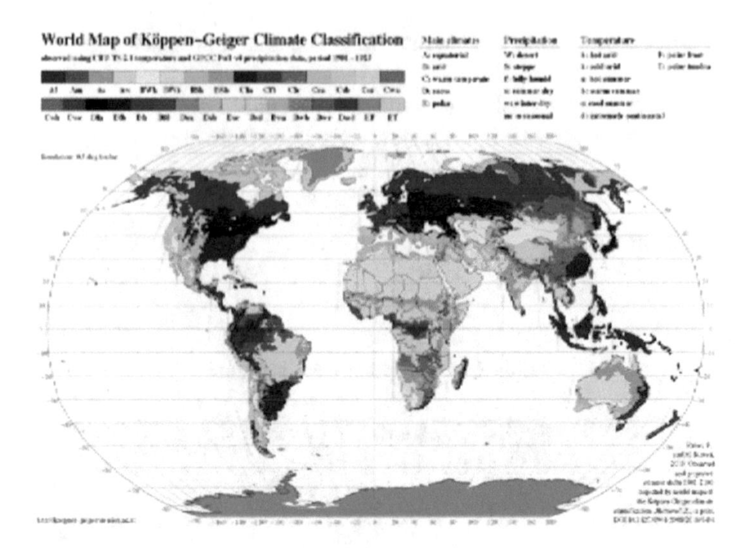

Fig. 1. Type of geographical climate according to the Köppen classification

Table 2. Description of the climate variable used in the koppen classification

Variable	Description
T	Air temperature
P	Precipitation
Subscript	
min	Minimum monthly value for the whole year
max	Maximum monthly value for the whole year
ann	Annual mean value
th	Threshold value[1]
mon	Number of months satisfying the criterion[2]
smin	Minimum monthly value for the summer months[3]
wmin	Minimum monthly value for the winter months[3]
smax	Maximum monthly value for the summer months[3]
wmax	Maximum monthly value for the winter months[3]
s	Mean value for the summer months[3]
w	Mean value for the winter months[3]

Table 3. The third letter

Description	Criterion
Hot arid	$T_{ann} \geq +18$ °C
Cold arid	$T_{ann} < +18$ °C
Hot summer	$T_{max} \geq +22$ °C
Warm summer	$T_{max} < +22$ °C, 4 $T_{mon} \geq +10$ °C
Cool summer	$T_{max} < +22$ °C, 4 $T_{mon} < +10$ °C, $T_{min} > -38$ °C
Cold summer	$T_{max} < +22$ °C, 4 $T_{mon} < +10$ °C, $T_{min} \leq -38$ °C

Table 4. The first and second letter

Type	Description	Criterion
A	Tropical climates	$T_{min} \geq +18$ °C
Af	Tropical rain forest	$P_{min} \geq 60$ mm
Am	Tropical monsoon	$P_{ann} \geq 25(100 - P_{min})$ mm
As	Tropical savannah with dry summer	$P_{min} < 60$ mm in summer
Aw	Tropical savannah with dry winter	$P_{min} < 60$ mm in winter
B	Dry climates	$P_{ann} < 10\ P_{th}$
BW	Desert (arid)	$P_{ann} \leq 5\ P_{th}$
BS	Steppe (semi-arid)	$P_{ann} > 5\ P_{th}$
C	Mild temperate	-3 °C $< T_{min} < +18$ °C
Cs	Mild temperate with dry summer	$P_{smin} < P_{wmin}$, $P_{wmax} > 3\ P_{smin}$, $P_{smin} < 40$ mm
Cw	Mild temperate with dry winter	$P_{smax} > 10\ P_{wmin}$, $P_{wmin} < P_{smin}$
Cf	Mild temperate, fully humid	Not Cs or Cw
D	Snow	$T_{min} \leq -3$ °C
Ds	Snow with dry summer	$P_{smin} < P_{wmin}$, $P_{wmax} > 3\ P_{smin}$, $P_{smin} < 40$ mm
Dw	Snow with dry winter	$P_{smax} > 10\ P_{wmin}$, $P_{wmin} < P_{smin}$
Df	Snow, fully humid	Not Ds or Dw
E	Polar	$T_{max} < +10$ °C
ET	Tundra	$T_{max} \geq 0$ °C
EF	Frost	$T_{max} < 0$ °C

Hence P is the amount of precipitation in millimetres and T is the temperature is in degrees Celsius.

This criterion aridity of Gaussen allows locating the places where Mediterranean vegetation will develop.

In the sense of Gaussen, the climate is [15]

- Xerothermic Mediterranean in the presence of 7 to 8 dry months.
- The climate is called thermo Mediterranean in the presence of 5 to 6 months dry.
- The climate is called Meson Mediterranean in the presence of 3 to 4 dry months.
- The climate is called sub-Mediterranean if the place has 1 or 2 dry months7 (Fig. 2).

3.4 Example of Rules of Koppen Use in Our Analysis

- Precipitation <40 mm during the driest month
- Precipitation of the driest month <1/3 of the wateriest winter month
- 3 °C <= Temperature of the coldest month >= 18 °C
- Climate is considered hot (CSA) if average temperature of the hottest month >22 °C
- Temperate climate (CSB) if average temperature of the hottest month <22 °C and has 4 consecutive months where the average is greater than 10 °C [15].

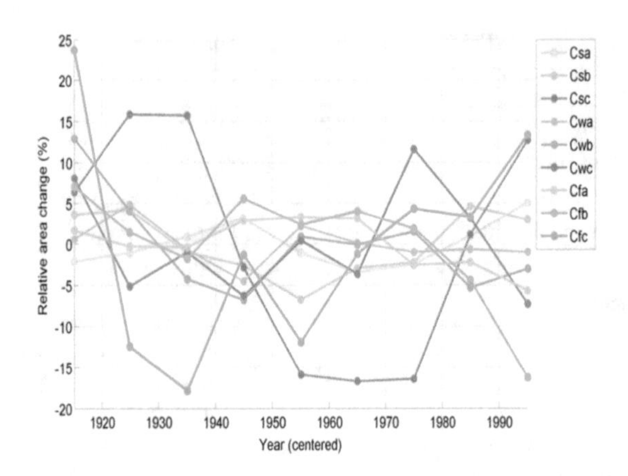

Fig. 2. It represents an example from our dataset table (test data that has been transformed Problems caused by climate change linked to the rule of Koppen in the Mediterranean Climate [15])

3.5 Example Rules of Gaussen Use in Our Analysis

– Gaussen aridity index: Henri Gaussen defines a dry month as a month where: $P < 2 \times T$.

4 Building a New Knowledge Data Based on the Data Mining Algorithm and the Climate Rule

4.1 Collecting Data

Dataset to Build the Model (Generating a New Database)
According to the Koppen Rules [4] we will apply these rules on the current monthly climate data to generate a new database containing the indices from the precipitation and the monthly temperatures we will classify if the year and wet or not if it Has been of the Mediterranean type or if the climate change has impacts on the climate to be continued to classify the years of the new data base using the algorithm of data mining such of the decision trees for the classification followed by neural network for the predictions.

Test Data
To test our model, the data are collected from NOAA, National Environmental Information Centres (NCEI). Is a weather and climate centre that offers more than 25 petabytes of complete atmospheric, coastal, oceanic and geophysical data. We used python as a language of programming [18].

As we see in this data, we extract information's that we call features that can only have 3 discrete values −1, 0 or 1, the bellow table explains it well, all the attributes having a binary values space are generally denoting the absence or presence of respective attribute. Attributes with three possible values are generally representing the strength: [17] (Fig. 3 and Table 5).

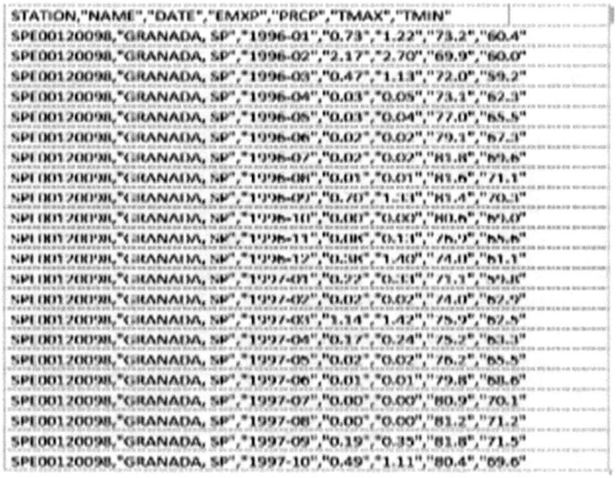
```
STATION,"NAME","DATE","EMXP","PRCP","TMAX","TMIN"
SPE00120098,"GRANADA, SP","1996-01","0.73","1.22","73.2","60.4"
SPE00120098,"GRANADA, SP","1996-02","2.17","2.70","69.9","60.0"
SPE00120098,"GRANADA, SP","1996-03","0.47","1.13","72.0","59.2"
SPE00120098,"GRANADA, SP","1996-04","0.03","0.05","73.1","62.3"
SPE00120098,"GRANADA, SP","1996-05","0.03","0.04","77.0","65.5"
SPE00120098,"GRANADA, SP","1996-06","0.02","0.02","79.1","67.3"
SPE00120098,"GRANADA, SP","1996-07","0.02","0.02","81.8","69.6"
SPE00120098,"GRANADA, SP","1996-08","0.01","0.01","81.6","71.1"
SPE00120098,"GRANADA, SP","1996-09","0.70","1.[illegible]","81.4","70.3"
SPE00120098,"GRANADA, SP","1996-10","[illegible]","[illegible]","80.6","[illegible]"
SPE00120098,"GRANADA, SP","1996-11","[illegible]","0.1[illegible]","76.9","65.6"
SPE00120098,"GRANADA, SP","1996-12","[illegible]","1.40","74.0","61.1"
SPE00120098,"GRANADA, SP","1997-01","0.22","[illegible]","71.1","[illegible]"
SPE00120098,"GRANADA, SP","1997-02","0.02","0.02","74.0","62.9"
SPE00120098,"GRANADA, SP","1997-03","1.14","1.42","75.9","62.5"
SPE00120098,"GRANADA, SP","1997-04","0.17","0.24","75.2","63.3"
SPE00120098,"GRANADA, SP","1997-05","0.02","0.02","76.2","65.5"
SPE00120098,"GRANADA, SP","1997-06","0.01","0.01","79.8","68.6"
SPE00120098,"GRANADA, SP","1997-07","0.00","0.00","80.9","70.1"
SPE00120098,"GRANADA, SP","1997-08","0.00","0.00","81.2","71.2"
SPE00120098,"GRANADA, SP","1997-09","0.19","0.35","81.8","71.5"
SPE00120098,"GRANADA, SP","1997-10","0.49","1.11","80.4","69.6"
```

Fig. 3. NOAA climate data

Table 5. Our new climate dataset, based on real rules of climate adapted to the cultivation of olives (−1 climate change, 0 for suspicious and 1 for good climate)

Features Group	Climate factor indicator	Values	Feature in our dataset
(1) Gaussen rules	Gaussen aridity index	−1, 1	G1
	Places where a Mediterranean vegetation will develop	1, 2, 3, 4, 5	G1
(2) Koppen rules	Precipitation <40 mm during the driest month	−1, 1	K1
	Precipitation of the driest month <1/3 of the wateriest winter month	−1, 1	K2
	3 °C <= Temperature of the coldest month >= 18 °C	−1, 1	K3
	Climate is considered hot (CSA) if average temperature of the hottest month >22 °C	−1, 1	K4
	Temperate climate (CSB) if average temperature of the hottest month <22 °C	−1, 1	K5
	4 consecutive months where the average is greater than 10 °C	−1, 1	K6
Result	Final result	−1, 0, 1	Prodc

5 Result

5.1 Data Mining Algorithms and Techniques for the Predictive Analysis Tested

Data mining and a data analysis tool allows the exploration of data according to a process of extraction of important and useful information from large sets, to arrive thereafter presented in a summarized, organized and analysis. The aim of datamining and thus repartitioning.

Table 2: Results of different classification algorithms built on our model. Note that, our dataset contains the history of 20 years of climate data input of real active.

On four simple techniques: prediction, identification, classification and optimization [6]. Beginning with the identification this data mining function serves to identify and make the relationship between the data. While the predictive model has the role of finding a model that identifies data classes or the concept whose data is organized, the objective being to be able to use this model to predict the label class Whose class is unknown [2]. While the classification and clustering technique is designed for classification Unknown samples using information provided by a set of Samples classified. This set is generally regarded as a Learning set. If a learning set is not available, there is no Prerequisite Knowledge on the data to be classified. In that case, grouping techniques were used to divide a set of unknowns in clusters [1] (Table 6).

Convert data to binary features: the input data point to be classified at left, and the predicted output (good olive production or bad) at right with some machine learning classifier (Table 7).

The test idea to compare the output of olive trees in Morocco by taking into account the climatic parameters and the rules of gaussen and koppen, each line is the line corresponding to a year our test history begins from 1996 and Ending in 2016 that we already have binary functionality as in Features in figure without target function, and prediction is the predicted value for each line).

Thus, if the prediction is less than 0, we will say that the climate does not correspond to the rules of gaussen and koppen d where the production will be ingerieure, else it is a good production [8]. In general, machine learning classifiers derive these performance results:

As expected, the decision tree gave us the best results, the decision tree is built on the entire dataset using all feature sets, Random Forest also gave good results Because it is an example of random forests, only a subset of rows is used randomly and many features are randomly selected to agree and a decision tree is built on that subset. SVM shows that it is also a good classifier (Fig. 4).

Table 6.

```
year,k1,k2,k3,k4,k5,g2,production
1996,1,-1,1,1,-1,2,-1
1997,1,-1,1,1,-1,2,0
1998,1,-1,1,1,-1,5,-1
1999,1,-1,1,1,-1,2,-1
2000,1,-1,1,1,-1,2,-1
2001,1,-1,1,1,-1,2,1
2002,1,-1,1,1,-1,1,-1
2003,1,-1,1,1,-1,2,1
2004,1,-1,1,1,-1,1,-1
2005,1,-1,-1,1,-1,1,-1
2006,1,-1,1,1,-1,1,1
2007,1,-1,1,1,-1,1,1
2008,1,-1,1,1,-1,2,0
2009,1,-1,1,1,-1,2,1
2010,1,-1,1,1,-1,1,1
2011,1,-1,1,1,-1,2,1
2012,1,-1,-1,1,-1,2,-1
2013,1,-1,1,1,-1,1,1
2014,1,-1,1,1,-1,1,-1
2015,1,-1,1,1,-1,2,1
2016,1,-1,1,1,-1,1,1
```

Table 7. Comparatives studies of the datamining algorithms

	Precision-prediction	Accuracy	Prediction test for 2017
Kernel-SVM	85%	0.727273	−1
Decision tree	99%	0.363636	−1
Random forest	97%	0.181818	−1
Logistic-regression	80%	0.714286	−1

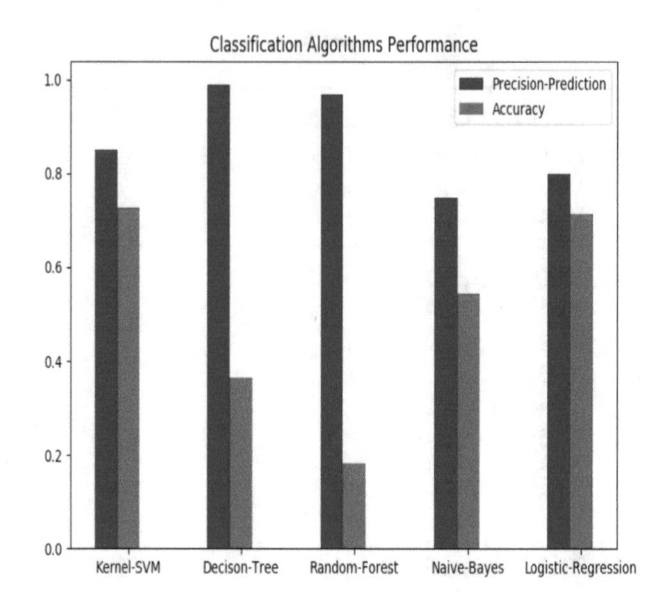

Fig. 4. The result of classification algorithms

6 Conclusion

In this article, the goal is to create a new technique for predicting agricultural yields. The idea is to rely on simple climate and statistical rules as well as historical and meteorological forecasts to build a new knowledge base reliable and parameterizable according to the season and culture. The idea is also to carry out a structural analysis that consists in extracting rules of normal climate and applying them to climate having a change of climate to compare this change with the climatic rules adapted for a better crop yield future. The results of this analysis were already presented in my first work [17]. The aim of this article was more oriented towards the explanation of the methodology and the rules of climate.

The future work will be the use of deep learning to explore the results and to compare the precision of the results from the result obtained before.

References

1. Patel, H., Patel, D.: A brief survey of data mining techniques applied to agricultural data. Int. J. Comput. Appl. **95**(9), 6–8 (2014). https://doi.org/10.5120/16620-6472
2. Mucherino, A., Papajorgji, P., Pardalos, P.: Survey of data mining techniques applied to agriculture. Oper. Res. **9**(2), 121–140 (2009). https://doi.org/10.17148/IJARCCE.2016.5263
3. Gupta, E.: Process mining a comparative study. Int. J. Adv. Res. Comput. Commun. Eng. **3** (11), 17–23 (2014). https://doi.org/10.17148/ijarcce
4. Koppen, C.S.: Climat Climat sec, 1–3 (1961)

5. Platt, J.C.: Sequential minimal optimization: a fast algorithm for training support vector machines. In: Advances in Kernel Methods, pp. 185–208 (1998). http://doi.org/10.1.1.43.4376 Sur, R., Aride, Z. (n.d.) Carte de la
6. Big Data et machine learning.pdf. (n.d.)
7. Huiles, D.: Olive - Olive Oils, November 2016
8. Olaiya, F.: Application of data mining techniques in weather prediction and climate change studies. Int. J. Inf. Eng. Electron. Bus. **4**, 51–59 (2012). https://doi.org/10.5815/ijieeb.2012.01.07
9. Chapman, L., Thornes, J.E.: The use of geographical information systems in climatology and meteorology. Progress Phys. Geogr. **27**(3), 313–330 (2003)
10. Iglesias, C., Torres, J.M., Nieto, P.J.G.: Turbidity prediction in a river basin by using artificial neural networks: a case study in Northern Spain. Water Resour. Manag. **28**, 319–331 (2014). https://doi.org/10.1007/s11269-013-0487-9
11. Goyal, M.K., Burn, D.H., Ojha, C.S.P.: Evaluation of machine learning tools as a statistical downscaling tool: temperatures projections for multi-stations for Thames River Basin, Canada. Theor. Appl. Climatol. **104**, 519–534 (2012). https://doi.org/10.1007/s00704-011-0546-1
12. Calzadilla, A., Zhu, T., Rehdanz, K.: Climate change and agriculture: Impacts and adaptation options in South Africa. Water Resour. Econ. **5**, 1–25 (2014). https://doi.org/10.1016/j.wre.2014.03.001
13. Luo, Q., Yu, Q.: Developing higher resolution climate change scenarios for agricultural risk assessment: progress, challenges and prospects. Int. J. Biometeorol. **56**, 557–568 (2012). https://doi.org/10.1007/s00484-011-0488-4
14. Ahmed, K., Shahid, S., Haroon, S.B., Xiao-Jun, W.: Multilayer perceptron neural network for downscaling rainfall in arid region: a case study of Baluchistan. J. Earth Syst. Sci. **6**, 1325–1341 (2015)
15. http://hanschen.org/koppen/
16. https://www.finances.gov.ma/Docs/depf/2018/summary_ref_plf2018.pdf
17. Fathi, M.T., Ezziyyani, M., Cherrat, L., Sendra, S., Lloret, J.: The relevant data mining algorithm for predicting the quality of production of olive in granada region influenced by the climate change, pp. 1–6 (2017). https://doi.org/10.1145/3175628.3175649
18. http://www.noaa.gov/

The Effect of Inoculation by Indigenous Endomycorrhizal Fungi on the Tolerance of *Tetraclinis articulata* Vahl masters Plants to Water Stress

Amal El Khaddari[1,2](✉), Jalila Aoujdad[2], Younes Abbas[3], Abdenbi Zine El Abidine[4], Mohamed Ouajdi[2], Salwa El Antry[2], and Jamila Dahmani[1]

[1] Laboratory of Botany, Biotechnology and Plant Protection, Faculty of Sciences, Ibn Tofail University, Kenitra, Morocco
elkhaddaribiologiste@gmail.com
[2] Center of Forest Research, Rabat-Agdal, Morocco
[3] Polyvalent R&D Laboratory, Polydisciplinary Faculty, Sultan Moulay Slimane University, Beni-Mellal, Morocco
[4] National School Forestry of Engineers, Sale, Morocco

Abstract. In this study, we examined several aspects related to water stress tolerance of Thuya (*Tetraclinis articulata* Vahl masters) inoculated with a native Arbuscular Mycorrhizal Fungi (AMF). The mycorrhizal and non-mycorrhizal Thuya were subjected to two water levels: under well-watered and under drought stress. Our results show that the AMF have a significant impact on biomass growth and the leaf water potential was also higher in stressed mycorrhizal plants (–18 bar), than in non-mycorrhizal plants (–41 bar). These AMF have also affected the water content, they have higher values compared to the control after 19 days of stress. These results confirm that inoculation with AMF can improve plant drought tolerance by increasing leaf water potential in mycorrhizal plants during drought, and can also provide a potential solution for the conservation and recovery of T. articulata in plants on natural ecosystems in Morocco.

Keywords: Water stress · Arbuscular Mycorrhizal Fungi · Water stress tolerance · *Tetraclinis articulata* Vahl masters · Morocco

1 Introduction

Tetraclinis articulata (Vahl) Masters is a resinous species that belong to the family of Cuprecessaeae. It is endemic to the western Mediterranean [1, 2] and almost exclusively North African [3]. *Tetraclinis articulata* is a great interest on the one hand in the protection of soil against erosion, and on the other hand as means of people's livelihood who consider it a valuable species in his wood very appreciated in the work of marquetry and the fabrication of decorative objects, crafts activities developed around this species generate considerable income for local people [4, 5]. Unfortunately, this

M. Ezziyyani (Ed.): AI2SD 2018, AISC 911, pp. 80–87, 2019.
https://doi.org/10.1007/978-3-030-11878-5_9

Moroccan forest ecosystem is confronted since the last decades the important degradation due to the expansion of anthropogenic activities [6], the biotic constraints (parasitic attacks, fungal diseases, insects, etc.) and abiotic stress (drought, erosion, fires and increased salinity). According to [7] the drought can be considered a major environmental constraint imposes effects on growth and development of plants and has one of the most important abiotic factors limiting growth and the yield of plants in many areas [8]. Multiple research using beneficial microorganisms of the soil, such as mycorrhizal fungi, have been tested to mitigate drought stress. Some of these studies [9] showed that inoculation by Arbuscular Mycorrhizal Fungi (AMF) can improve the absorption of water in many plant species under water stress and can protect the host plants against drought. In fact, mycorrhizal colonization can also directly improve the water relations of plants by increasing the absorption of roots, leaf water potential, by regulating the rate of transpiration [10], and increasing the rate of photosynthesis [11]. According to [12] shows at drought period, the osmotic adjustment is important in mycorrhizal leaves. Likewise [13] shows that in the plants inoculated by AMF, could delay the decline of leaf water potential (Ψ) during water stress. However according to [14] have found when the water was not limiting, the water potential of the leaves (Ψ) was similar in plants inoculated and plants not inoculated by AMF. The aim of this work is to study the role and the effect of the controlled mycorrhization on drought tolerance of Thuya plants in areas subject to drought. This study revealed on the morphological parameters and water relations in the shoot and roots of T. articulata. We will measure the basic and midday water leaf potential, as well as the shoot and root dry mass, the water content in the leaves and soil water content.

2 Materials and Methods

2.1 Plant and Fungal Material

The Plants of Thuya used in this study come from seeds from the Azrou Seed Center. Half of these seeds were planted in soil inoculated with indigenous AMF inoculum isolated from the rhizosphere of Beni Souhane tetraclinaie located in the Middle Eastern Atlas, Morocco. The other half was planted in a soil not inoculated with Mycorrhizal Fungi called Non-AMF, it is a control. The AMF and Non-AMF Thuya seedlings were grown at the nursery of Beni Souhane in natural light with daily watering between April 2016 and November 2016. After 8 months of stay in the nursery of Beni souhane, the Forestry Research Center has brought back eight portoirs each portoirs contains 54 alveoli with a total of 432 Thuya plants: 4 portoirs mycorrhizal and 4 portoirs non-Mycorrhizal. The center has made these last at our disposal for effectuating a test of water stress, and proceed for the morphological parameters and water relations in the leaves.

2.2 The Experimental Design

The experiment was conducted at the nursery of the Forest Research Center (FRC) of Rabat (34° 01′ N, 6° 5′ W), located at 135 m altitudes under a bioclimatic of sub-humid at hot winter. The annual average rainfall is 555 mm and the average annual temperature is 17.9 °C [15, 16].

The portoirs were arranged according to a randomized complete block (two blocks) with two treatments: the first treatments: inoculation with AMF or no AMF (control) (Mycorrhizal and non-Mycorrhizal plant) and the second treatments: concerns the regime of watering: watering controlled (100% watered all along the experiment), severe stress regime (stop watering all along the experience). The Water stress lasted 19 days, from 01/08/2017 to 19/08/2017 on 2 portoirs of mycorrhizal plants and two portoirs of non-mycorrhizal plants.

During this experiment we took three samples for each treatment and for each date (T0, T1, T2 AND T3) with T0 = 07/08/2017; T1 = 10/08/2017; T2 = 15/08/2017 and T3 = 19/08/2017.

2.3 Parameters Measured

The response of T. articulata plants to mycorrhizal inoculation was estimated by determining the plant biomass production, leaf water potential, soil water content (W) and the water content of shoots (WC).

- **Biomass production**
  The shoot and root systems were separated and the shoot dry weight (DW) was measured after drying in an oven at 60 °C for 48 h.
- **Leaf water potential Ψ**
  The basic and midday water leaf potential (Ψ) was determined using a pressure chamber of Scholander. The needle of the Thuya leaf was isolated and fixed with soft tapes then squeezed into pores of rubber seals. This last was thereafter fastened, and the potential (Ψ) was recorded when there was continuous water efflux from the leaf base, depending on a pressure bottle to increase the pressure chamber.
- **Soil water content (W)**
  This parameter was estimated by taking a soil sample and determining wet weight (WW) and then determining the dry weight of the sample (DW) which was measured after drying in an oven at 60 °C for 48 h. The water content (denoted W) was calculated as follows:

$$\text{Soil water content (W)} = [(\text{weight of wet soil (WW)} - \text{weight of dry soil (DW)})/\text{weight of dry soil (DW)})] \times 100.$$

2.4 Statistical Analysis

The data was analyzed with three-way ANOVA, using SPSS software version 20.0. We express the data on average ± standard deviation of the observations. The three-way ANOVA was enabling on the data of all the parameters studied to compare the effects of mycorrhizal inoculation, different levels of stress and duration of water stress in addition to the multiple-range Duncan test of $P < 0.05$ [17].

3 Results

3.1 Biomass Production

Interestingly, the results show that the effects of water stress, mycorrhizal inoculation and the time of stress has affected significantly the shoot and root dry weight of the plants of Thuya on the threshold of 5% (Fig. 1). Under well-watered (Fig. 1), the averages of the shoot dry weight of mycorrhizal and Non-mycorrhizal Thuya plants were almost similarly throughout the stress period. In mycorrhizal plants, the values are then: 2.33; 1.93; 3 and 3.83 g respectively to T0, T1, T2, and T3 and Non-mycorrhizal plants, which show values of the order of 3.76; 3.63; 4.23 and 5.4 g respectively to T0, T1, T2 and T3. Under water stress, the results showed that the shoot dry weight decreased highly significantly for Non-mycorrhizal plants with values: 1.96; 3.03; 3.53 and 3.8 g respectively to T0, T1, T2 and T3. Than in mycorrhizal plants that show levels of the order of: 1.66; 1.8; 2.16 and 2.63 g respectively to T0, T1, T2 and T3.

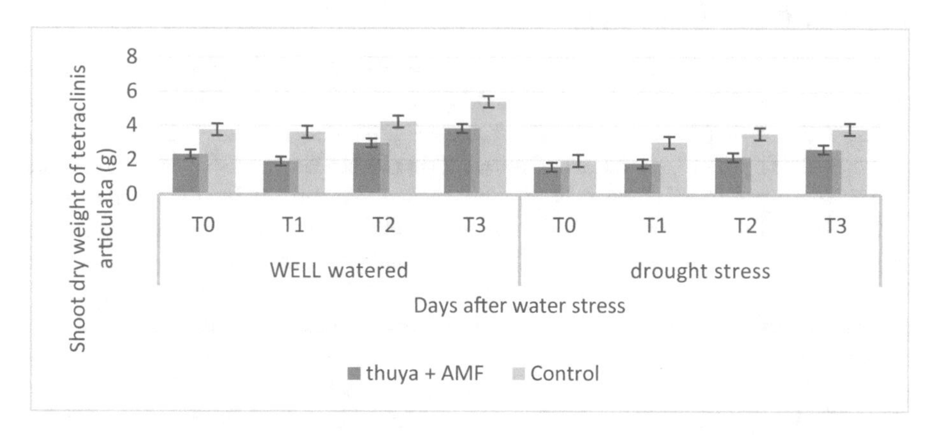

Fig. 1. Shoot dry weight of plants T. articulata in mycorrhizal (M) and non-mycorrhizal (NM) plants under water stress and well-watered.

The averages of the root dry weight in well-watered of the mycorrhizal and Non-mycorrhizal plants of Thuya are similar (Fig. 2) all along the period of stress. In mycorrhizal plants, the values are then: 3.1; 3.1; 3.2 and 3.76 g, respectively to T0, T1, T2, and T3 and Non-mycorrhizal plants, which show the values of the order of 4.13; 4.03; 4.36 and 4.66 g respectively to T0, T1, T2 and T3. Under water stress, the results showed that the root dry weight decreased highly significantly for mycorrhizal plants with values: 3.11; 2.63; 1.73 and 1.93 g respectively to T0, T1, T2 and T3. As Non-mycorrhizal plants that show levels of the order of 1.36; 3.46; 3.43 and 2.9 g, respectively to T0, T1, T2 and T3.

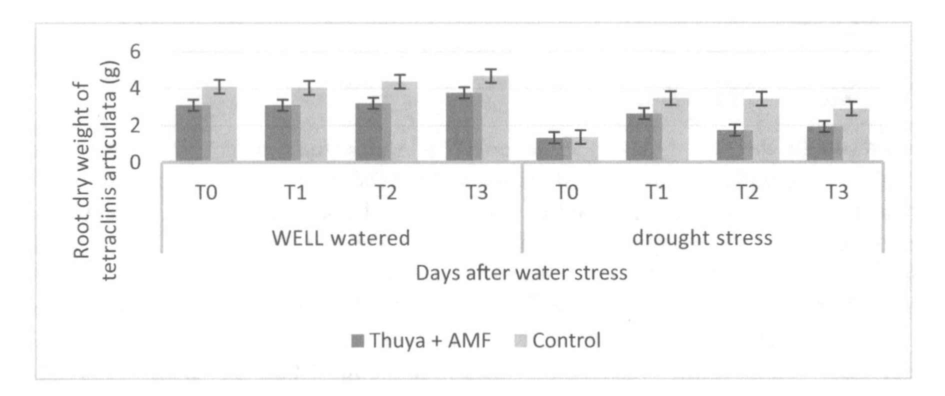

Fig. 2. Root dry weight of plants T. articulata in mycorrhizal (M) and non-mycorrhizal (NM) plants under water stress and well-watered.

3.2 Leaf Water Potential (Ψ)

The results show that the effects of water stress, mycorrhizal inoculation and stress time have significantly affected the basal and midday water potential (Ψ) of the plants of Thuya in the 5% threshold (Figs. 3 and 4). The results (Fig. 3) show that the mycorrhizal plants of T. articulata have a higher basal leaf water (Ψ) in needles between –6.5 bar and –13 bar than in Non-mycorrhizal plants in needles between –6 bar and –21 bar subjected to water stress. While Non-mycorrhizal plants and mycorrhizal plants showed a basic water leaf (Ψ) is not significant in needles between (–0.5 bar and –7.6 bar) under well-watered. The Fig. 4 show that after 19 days of water stress, the midday water leaf Ψ reached a maximum values of the order of –40 bar in Non-mycorrhizal plants of T. articulata against the mycorrhizal plants is reached mean values of midday water leaf (Ψ) in the order of –18 bar. While Non-mycorrhizal plants and mycorrhizal plants showed a midday leaf water (Ψ) not significant in needles between (–2 bar and –9 bar) under well-watered.

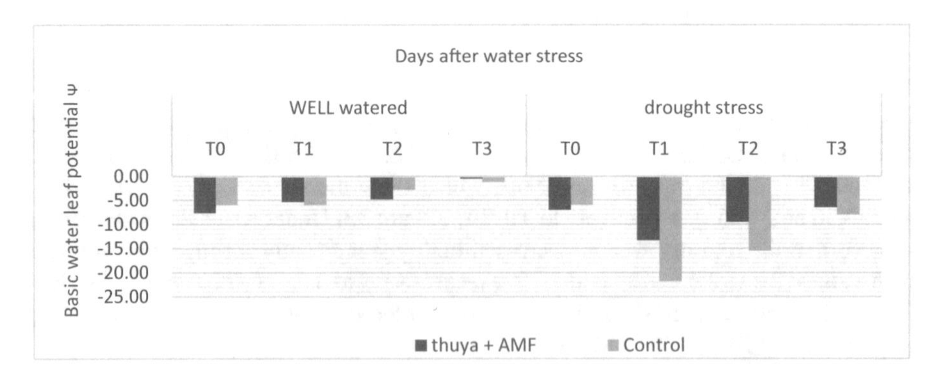

Fig. 3. Basic water leaf (Ψ) of plants T. articulata in mycorrhizal (M) and non-mycorrhizal (NM) plants under water stress and well-watered.

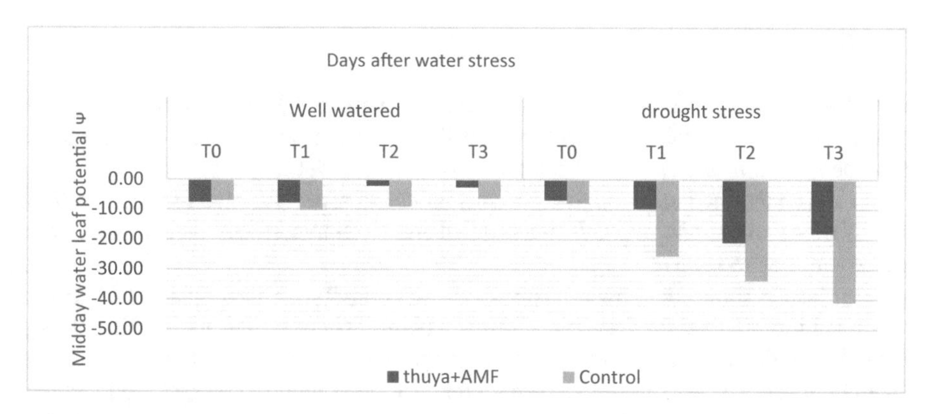

Fig. 4. Midday water leaf (Ψ) of plants T. articulata in mycorrhizal (M) and non-mycorrhizal (NM) plants under water stress and well-watered.

The soil water content was significantly affected by the water stress, mycorrhizal inoculation and the time of stress in the threshold 5% (Fig. 5). Under well-watered water, the average soil water content of mycorrhizal and Non-mycorrhizal Thuya plants was almost similar. Under water stress, the results have shown that the soil water content increased highly significantly for mycorrhizal plants with the values are in order of 17.36; 11.52; 15.79 and 14.63 (%) respectively to T0, T1, T2 and T3. Likewise for Non-mycorrhizal plant the values are in order of 14.34; 9.29; 14.3 and 13.19 (%) respectively to T0, T1, T2 and T3.

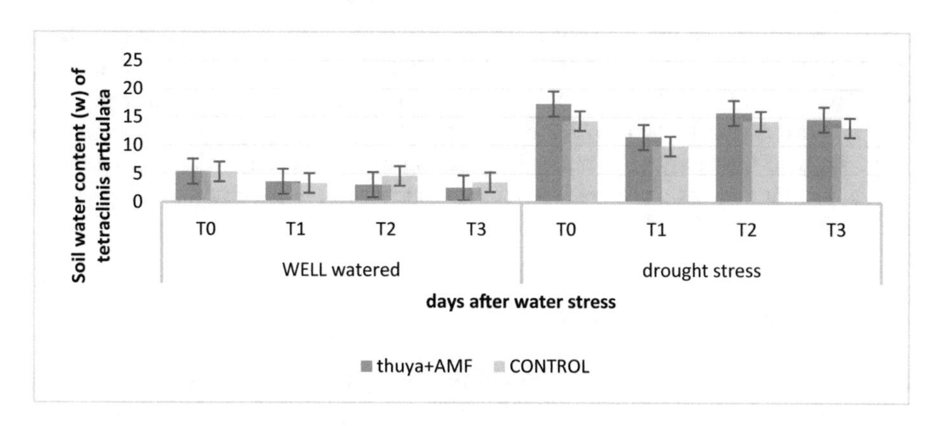

Fig. 5. Soil water content (W) of plants T. articulata in mycorrhizal (M) and non-mycorrhizal (NM) plants under water stress and well-watered.

4 Discussion

The drought has profound effects on plant production [18] and to counteract this constraint, several studies have shown that the arbuscular mycorrhizal symbiosis and absorbent hairs to increase plant tolerance to the water deficit [10].

In this study, the morphological and physiological aspects related to water relations and tolerance to drought in plants mycorrhizal and Non-mycorrhizal plants subjected to water stress were studied.

Our results shows that, under water stress, the non-mycorrhizal plants had a higher shoot and root dry weight than in mycorrhizal plants. These results are in contradiction with the studies carried out by [19, 20] on the plants of Cupressus atlantica G and Quercus suber. These authors have shown that dry shoot and root weight production in mycorrhizal plants is higher than in non-mycorrhizal plants. For our results about the shoots and root, in mycorrhizal conditions, according to our results the mycorrhization can cause a reduction in growth these results are in agreement with [21]. It has also been demonstrated that a latency period allowing an inoculated fungus to settle can reduce the initial growth gain of the host relative to non-inoculated ones.

About the plants water status water potential and relative humidity) the inoculation of thuya plants by native endomycorrhizal allows of the plant to obtain a content of water and the leaf water potential important under drought stress. These results are agree with [19, 20]. These authors showed that mycorrhizal C. atlantica plants and Quercus suber have higher water potential and water content than control plants under water stress. Several studies [22] reported that water content is considered a good indicator of water stress tolerance in a wide variety of plants. Otherwise, the Physiological responses of mycorrhizal symbiosis may depend on the host plant and fungal species. However, the influence of these responses on leaf water potential in mycorrhizal plants under water stress still unclear. Several studies have shown that under water stress, mycorrhizal plants with a water deficit maintain a higher water potential than non-mycorrhizal plants [23]. Preserving higher water content in mycorrhizal plants suggests a strategy of tolerance by reducing water loss and avoiding the pressure exerted by water stress on the plant. Similar responses have been reported for several mycorrhizal plants [23–25].

5 Conclusion

In conclusion, the present results showed that drought stress could cause a positive impact on plant growth. The results of our study suggest that the inoculation of T. articulata with AMF improve plant tolerance to drought stress. By improving plant water relations, as indicated by increased leaf water potential and soil water content. Which could be explained by an improved water uptake directly by extended extraradical hyphae of AMF. In summary, our studies clearly demonstrated that indigenous AMF were effective in improving the tolerance of T. articulata under water stress.

References

1. Rikli, M.: Das Planzenkleid der Mittellmeerlander. Huber Berne **1**, 1–41 (1943)
2. Quezel, P.: Biogéographique et écologie des conifères sur le pourtour Mediterranean. Impression: Actualité d'écologie forestière, pp. 205–256. Badras Edit, Paris (1980)
3. Quezel, P.: Réflexion sur l'évolution de la flore et de la végétation, au Maghreb Méditerranéen, p. 117. Ibis Press Edit, Paris (2000)
4. Bellakhadar, J.: La pharmacopée marocaine traditionnelle. Impressions DUMAS, Saint-Etienne (1997)
5. Boudy, P.: Guide du forestier en Afrique du Nord, vol. 273. La maison rustiqu, Paris (1952)
6. Haut commisseriat des eaux et forêts, Le thuya de Berbérie. http://www.eauxetforets.gov.ma/fr/text.aspx?id=1061&uid=74. Accessed 12 Oct 2017
7. Denby, K.: Engineering drought and salinity tolerance in plants: lessons from genome-wide expression profiling in Arabidopsis. Trends Biotechnol. **23**, 547–552 (2005)
8. Kramer, P.: Water Relations of Plants and Soils. Academic Press, San Diego (1995)
9. Ruiz-Lozano, J.M.: Arbuscular mycorrhizal symbiosis and alleviation of osmotic stress: new perspectives for molecular studies. Mycorrhiza **13**, 309–317 (2003)
10. Augé, R.M.: Foliar dehydration tolerance of mycorrhizal cowpea, soybean and bush bean. New Phytol. **151**, 535–541 (2001)
11. Augé, R.M.: Mycorrhizal promotion of host stomatal conductance in relation to irradiance and temperature. Mycorrhiza **14**, 85–92 (2004)
12. Augé, R.M.: Water relations, drought and vesicular-arbuscular mycorrhizal symbiosis. Mycorrhiza **11**, 3–42 (2001)
13. Subramanian, K.S.: Influence of arbuscular mycorrhizae on the metabolism of maize under drought stress. Mycorrhiza **5**, 273–278 (1995)
14. Kubikova, E.: Mycorrhizal impact on osmotic adjustment in Ocimum basilicum during a lethal drying episode. J. Plant Physiol. **158**, 1227–1230 (2001)
15. Les coordonnées géographiques de Rabat, Maroc. http://www.dateandtime.info/fr/. Accessed 09 May 2018
16. Climat: Rabat. https://fr.climate-data.org/location/3669/. Accessed 09 May 2018
17. Duncan, D.B.: Multiple range and multiple F-tests. Biometrics **11**, 1–42 (1955)
18. Jones, H.G.: Monitoring plant and soil water status: established and novel methods revisited and their relevance to studies of drought tolerance. J. Exp. Bot. **58**, 119–130 (2007)
19. Zarik, L.: Use of arbuscular mycorrhizal fungi to improve the drought tolerance of Cupressus atlantica G. Comptes Rendus Biologies **339**(5–6), 185–196 (2016)
20. Nasslahsen, B.: Physiological responses of Quercus suber to the effect of water deficit in presence of ectomycorrhizal fungi. J. Appl. Phys. Sci. **4**(1), 33–45 (2018)
21. Orcutt, D.: The Physiology of Plants Under Stress: Soil and Biotic Factors. Wiley, New York (2000)
22. Shaw, B.: Responses of sugar beet (Beta vulgaris L.) to drought and nutrient deficiency stress. Plant Growth Regul. **37**, 77–83 (2002)
23. Meddich, A.: Rôle des champignons mycorhiziens à arbuscules de zones arides dans la résistance du trèfle (Trifolium alexandrinum L.) au déficit hydrique. Agronomie **20**(3), 283–295 (2000)
24. Gemma, J.N.: Mycorrhizal fungi improve drought resistance in creeping bentgrass. J. Turfgrass Sci. **3**, 15–29 (1997)
25. Ruiz-Lozano, J.M.: Alleviation of salt stress by arbuscular-mycorrhizal Glomus species in Lactuca sativa plants. Physiol. Plan. **98**, 767–772 (1996)

AgriFuture: A New Theory of Change Approach to Building Climate-Resilient Agriculture

Hajar Mousannif(✉) and Jihad Zahir

LISI Laboratory, Cadi Ayyad University, Marrakesh, Morocco
{mousannif, j.zahir}@uca.ac.ma

Abstract. Agriculture in Morocco, like in many developing countries, remains very sensitive to climatic fluctuations, with drought occurring recurrently; creating volatility in agricultural production and impacting negatively the lives of farmers. How to quantify the impact of climate change on the quality of life of farmers? How can climate-resilience be strengthened and livelihoods of farmers enhanced? How to make the adoption of improved agricultural technologies and practices by farmers sustainable? This paper aims at answering all those questions by presenting a new Theory of Change approach targeting the construction of comprehensive and large-scale datasets which integrate data from a wide range of stakeholders. Advanced data analytics will be applied on those data to provide a thorough understanding of the interrelated climatic, environmental, social, cultural, economic, institutional and political factors that aggravate differentiated climate change impacts. This will allow discovering hidden patterns in the data, making decisions and establishing recommendation systems guiding stakeholders' choices in terms of policies, irrigation decisions, types of crops to plant, and actions to take to enhance crop yield production, in order to make the most vulnerable communities more resilient to climate change.

Keywords: Agriculture · Theory of Change · Big Data Analytics · Machine learning · Climate change

1 Introduction

In Morocco, agricultural research has made much contribution to poverty reduction and food security over the last 30 years. However, it is noticeable that the awaited global change remains far from what was expected for many reasons. First, there is a striking dichotomy between large and small farms: more than 70% of farmers work fewer than five hectares, but this accounts for only a quarter of the total land under cultivation; the large farms dominate the fertile areas and have a more substantial income, earning approximately nine times more than the average family farm [1]. Moreover, Morocco's rural areas have poor socio-economic infrastructure, low levels of education and an ageing farmer population. Small farmers are economically vulnerable, particularly to climate change. Most of agricultural research initiatives conducted in Morocco so far have come up with perceived technology with little consultation with the farmers as the end users of the product. Technologies generated from such research often lack

M. Ezziyyani (Ed.): AI2SD 2018, AISC 911, pp. 88–97, 2019.
https://doi.org/10.1007/978-3-030-11878-5_10

relevance to farmers (especially small ones) and their adoption is usually very limited despite their demonstrated benefits. Moroccan smallholders have a huge wealth of knowledge of their production system as well as how they cope with climate change. However, such knowledge is not tapped along the process of developing adaptable technologies. Other factors also hinder the adoption of modern technologies, policies and practices. Institutional constraints such as non-availability of fertilizers and agro-chemicals during the growing season are a serious problem; because of climate change, roads especially to lands in mountainous regions are cut either because of exceptional heavy snow/rain or poor road infrastructure. Furthermore, private sector practitioners such as input dealers, transporters, processors, equipment hiring outfits, etc. are left out from the research/technology design despite their vital role. Indeed, policy makers fail in developing supportive policies and developing needed infrastructure because of the lack of communication and cooperation with the required stakeholders.

In this paper, we present a new Theory of Change Approach we refer to as AgriFuture and which allows the development of a robust and climate-resilient agriculture system to empower future farmers. The approach can be applied to Moroccan agriculture and enlarged to include any (developing) country. The proposed approach aims to ensure large and sustainable adoption of modern technologies/practices and policies and relies on the engagement of all necessary stakeholders on the value chain and also those that influence it from outside.

The remainder of this paper will be organized as follows: Sect. 2 gives an overview of projects targeting the empowerment of agriculture through modern technologies. Section 3 presents the proposed Theory of Change approach and its potential of achieving robust agriculture system which is resilient to climate change and climate hazards. It also focuses on the development outcomes fostered by the approach. Finally, Sect. 4 concludes the paper.

2 Related Work

Most initiatives targeting modern technologies development rely exclusively on the use of remote sensing (satellite imagery) and/or climatic parameters for crop yield monitoring, crop forecasting and water management.

CGMS-MAROC [2] (Comprehensive and Advanced Crop Monitoring System for Morocco) is an adaptation of Europe Crop Monitoring System (CGMS). It allows crop monitoring and yield forecasting in Morocco. Its deployment scale is very limited.

ACCAGRIMAG [3] is a project that aims to reduce the climate risk in agriculture. It targets rainfed agriculture with high cereal dominance. Crop losses are estimated from climatic parameters (or indices) or from satellite imagery, at the scale of communes.

IRRISAT [4] (Irrigation assisted by Satellite), aims to develop a system for irrigation based on the use of satellite imagery. The observation from space of agricultural land is used for monitoring the development of crops, making it possible to estimate the maximum amount of water to use for irrigation within hours of satellite acquisition.

IRRiEYE [5] aims at providing farms and managers of water resources with real time information on agricultural water needs. Irrigation needs are estimated using high resolution data from Earth observation satellite and standard international methodologies. Data are aggregated at various spatial scales (from field or irrigation unit to district or river basin scale) and temporal scales (real time, historical series). Information is directly sent by text (SMS) or by email; this can also be visualized through private access web-mapping applications based on Google Maps.

The drawbacks of existing systems targeting climate change resilience in agriculture are:

- They are mere pilot projects with very limited deployment scale.
- They focus more on technological aspects rather than the real need of farmers.
- They involve very few stakeholders.
- They do not take into account factors such as gender, ethnicity, socio-economic status, age and physical ability of farmers.
- They rely on very limited sources of data (satellite imagery or weather).
- They have a "local" nature and the results can hardly be tuned to include different contexts (different land covers, different quality of soil, different climatic conditions, etc.).

This paper attempts to overcome such drawbacks by presenting an approach that aims to achieve the following objectives:

- Provide a thorough understanding of the interrelated climatic, environmental, social, economic, and institutional factors that aggravate differentiated climate change impacts and guide better decision-making in the context of Moroccan agriculture and empowerment of marginalized farmers.
- Creation and integration of datasets built on real data derived from many stakeholders along with data from remote sensing and weather stations.
- Implementation of efficient data analytics, data mining and machine learning systems, and quantification of the effect of climate change on different value chains.
- Development of software interfaces for collaboration between different stakeholders allowing decision mapping, data visualization, and recommendation systems guiding decisions of both marginalized farmers and policy makers.

3 AgriFuture: A New Theory of Change Approach for Building Climate-Resilient Agriculture

Theory of Change (ToC) is the process of determining all the building blocks required to bring about a long-term goal. It describes the process of social change by making explicit the perception of the current situation; its underlying causes, the long term change desired and the things that need adjustment for the change to happen [6]. To achieve the aforementioned objectives presented in the previous section, the proposed ToC approach will be explained through the data analysis pipeline explained in [7, 8], from data collection to insights extraction with a focus on development outcomes.

3.1 Data Collection

In the proposed approach, different data sources can be used. They include:

- Climatic data: accessible from weather stations.
- Environmental data: relevant data related to agricultural lands, locations, nature of planted crops and soil quality. Remote sensing can be used to fill the gap of missing data, especially in mountainous regions or lands that are inaccessible.
- Social and cultural data: data related to communities including: gender, ethnicity, socio-economic status, age and physical ability.
- Economic data: data related to agricultural production per region and per agricultural season, as well as the Return on Investment.
- Transport infrastructure data: This data is particularly useful when exceptional rain/snow episodes occur. Roads can be cut for many days/weeks isolating marginalized agricultural areas and negatively affecting farmers (provisioning, marketing, etc.). This data represents archives of all roads cut in the past and the environmental conditions associated with the cuts.
- The farmers themselves as a source of data: data about the needs of farmers, their requirements and how they cope with climate-related hazards should be gathered.

3.2 Data Pre-processing

After data is collected, it is important to lay the ground for data analysis by applying various preprocessing operations to address potential imperfections in the raw data [9]. Data collection methods and conditions are not flawless; indeed missing values and data lack may hinder the process. Finally, data may simply need to fit requirements of analysis algorithms [10, 11]. Thus, collected data must be inspected, fused and all the above problems corrected during a pre-processing phase. Data quality management should serve as a basis for all the work, because of the importance of data quality in the whole decision-making process. Many data enhancement approaches, such as outlier detection, interpolation, data integration, data deduplication and data cleaning [12] should be investigated.

Considering that data are collected by heterogeneous sources and might manifest various anomalies, a middleware [13, 14] (e.g. Oracle Event Processing) can be used to abstract the different data preprocessing operations needed for each specific data source. A pipeline composed of various preprocessing stages will be implemented.

Each data source stream can be directed to an end-point where it undergoes all the stages needed to ensure their fitness for use in the system. Such stages include: (a) outlier detection [15], (b) interpolating missing values [16, 17], (c) data cleaning [18–20] (e.g. geometric correction, radiometric correction, etc.), and (d) aggregation and transformation [21]. Once all these stages passed, the various “cleaned” streams should be combined to form the final dataset used for data analysis.

3.3 Data Analysis

There exist many algorithms to perform various analytics on either structured or unstructured data. Four types of Analytics (descriptive, inquisitive, predictive and prescriptive) can be used in the proposed approach (as depicted in Fig. 1).

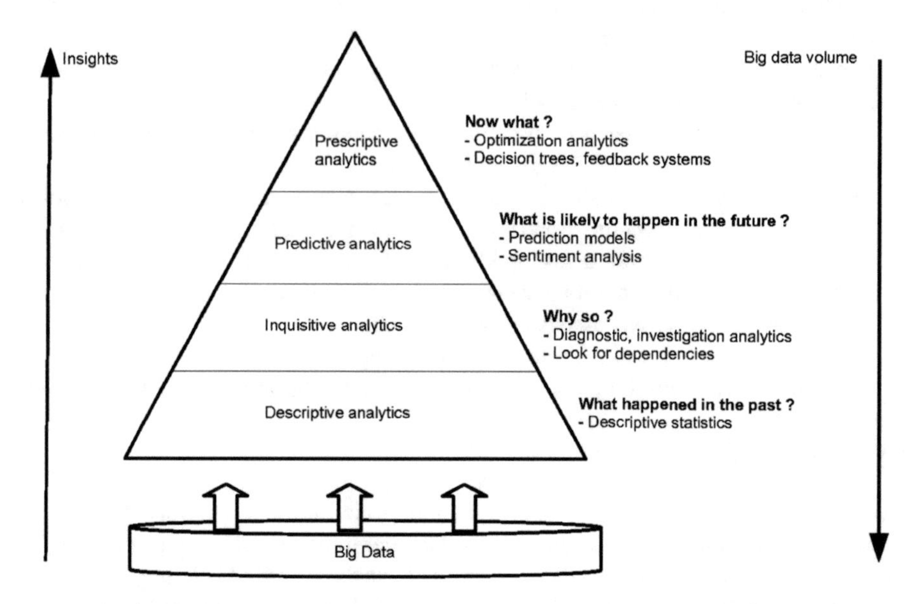

Fig. 1. Data analytics types [7]

Descriptive Analytics will allow understanding the data and answering questions such as: What is the average number of farmers working in mountainous areas? What is their Return on Investment? How many girls quit schools because of climate change? Etc. It uses descriptive statistics such as sums, counts, averages, min, and max to provide meaningful results about the analyzed dataset.

Inquisitive Analytics includes analytical drill downs into data, factor analysis, and conjoint analysis, among others. It will allow answering questions such as "why something is happening?" "What is the impact of gender on ROI?", "why Berbers are more resilient to climate change than Arabs?", "why the family income in certain regions is higher than that in other regions, given similar climate change hazards?"

By integrating different sources of data, many patterns in the data will emerge. The results will be far better compared to the case where only one source is used. Examples of findings may include: "When the average age of women is 25 years old, resilience to climate change is higher" or "When women participate in decision-making, ROI increases by x%".

Predictive Analytics consists of studying the data we have, to predict future outcomes in a probabilistic way. This could include:

- crop production estimates and crop growth predictions
- crop field and land cover classifications
- crop stress, and crop disease predictions
- Identification of appropriate crops to plant in fields based on previous successes and factors impacting crops growth.

Prescriptive analytics aims not only to predict, but to plan actions and provide recommendations for farmers, and decisions makers. By the use of advanced analytics and especially prescriptive analytics on gathered data, decision makers will not only be able to predict hazards and outcomes, but anticipate their occurrence and know in advance why they will happen. This will allow decision makers to be more prepared for what will be happening and adapt policies in advance. If governmental entities, for example, know from past experiences that the same roads will be cut during specific periods of time due to climate hazards and isolate vulnerable populations, why wait until this happens and struggle in finding palliative solutions to help reach those communities? Alternatives for basic needs storage/provisions should be set in advance and farmers need to be aware in advance that this will happen so they can get prepared. The strongest the relationship between different stakeholders is, the more resilient to climate change hazards communities will be.

Data analytics will provide a thorough understanding of the interrelated climatic, environmental, social, cultural, economic, and institutional factors that aggravate differentiated climate change impacts. The results from this understanding will lead to a reset in biased attitudes and mindsets and tackle discriminatory behaviors in agriculture. Policies based on this understanding will be focused on farmers' needs and this will create a shift in social equity. Indeed, by adopting the good practices, policies and recommendations established based on data analysis, agriculture should be able to provide everyone involved with a good quality of life and contribute to food sovereignty and reduction of poverty regardless of gender, ethnicity, social status or climate hazards. This will be achieved through the creation of synergies and effective collaboration at all levels and all parties including farmers, workers, processors, distributors, traders, consumers, public authorities, communities, and civil society.

3.4 Data Visualization

Visualization guides the analysis process and presents results in a meaningful way. For the simple depiction of data, most software packages support classical charts and dashboards. The choice is generally dictated by the type of desired analytics. There are challenges related to the volume such as processing time, memory limitations, and the need to fit different display types. Various techniques for data visualization can be used. Visual summaries, for instance, can present results for different stakeholders in a meaningful way and based on their needs.

A software interface that will serve as a platform of cooperation between different stakeholders including farmers should be considered. Such platform will allow farmers to access different types of analytics and make information about farm production (Type, quantity, quality, etc.) accessible online. This guarantees visibility of output and offers a fair competition between large and small farms.

3.5 Development Outcomes

The anticipated development outcomes for adopting the proposed ToC approach are:

Higher Adoption of Good Practices and Technologies: the proposed approach will raise awareness of farmers to technologies/practices that are tailored to their needs and are adaptable, economically viable and sustainable. Because the knowledge of farmers will be tapped along the design process, farmers will be more willing to adopt modern technologies.

Increased Quality and Quantity of Farm Productivity: The prediction models based on gathered multi-source data will lead to rationalization of the use of water for irrigation, will help in crop monitoring, crop disease predictions, drought prediction and will give insight on nature of crops to plant in yields based on previous successes and failures thanks to powerful machine learning algorithms.

Decreased Poverty and Increased Income for Farmers and Their Quality Of Life: the dichotomy between large and small farms will be less severe since small farms will gain as much visibility as large farms. Through the software interface the approach targets and to which all involved stakeholders (including farmers, exporters and SMEs) have access, the productivity, the quality and the type of crops will be displayed allowing a fair competition between small farms and large ones.

Resilience to Climate Change: because of the strong collaboration between all stakeholders (and especially between governmental agencies), policies will be set prior to harsh climatic episodes allowing vulnerable population to cope with climate hazards.

Empowerment of Women Farmers: Through data Analytics, the contribution of women will be statistically highlighted and their involvement in the decision-making process and its impact on development will be quantified in figures and not just words. This will confirm their role as active actors in this sector and open up new opportunities.

A pathway of change for the proposed approach is provided in Table 1.

Table 1. Pathway of change for AgriFuture

Approach Title: AgriFuture

Beneficiaries: Farmers and decision makers in agriculture

Necessity of the intervention

- Morocco's rural areas have poor socio-economic infrastructure, low levels of education and an ageing farmer population.
- Small farmers are economically vulnerable, particularly to climate change.
- The agricultural sector in Morocco is very sensitive to climatic fluctuations, with drought occurring every third year, creating volatility in agricultural production, and hence, affecting the whole family income.
- Young girls are forced to quit school to support their families in farming tasks.
- Women working in agriculture are usually unqualified and underpaid; the influence of climate change on agriculture, further exacerbates their exposure to poverty and vulnerability.

Approach Outcome / Outputs / Main Interventions

Outcome: Agricultural adaptation to climate change is improved in most exposed regions in Morocco

Output 1: Innovative systems to enhance resilience of farmers to climate change, in a selected region, are developed and used

– Collect and integrate relevant qualitative and quantitative data, including demographic, socio-economic, environmental and climatic data
– Implement efficient data analytics, data mining and machine learning systems, and quantification of the effect of climate change on different value chains

Output 2: Exposed farmers are sensitized to the need of new technologies to adapt with climate change

– Organize workshops to raise awareness of farmers regarding the new technologies
– Develop trainings and knowledge transfer programs adapted to farmers needs
– Support access of farmers to predictive analytics that are relevant to their agriculture
– Ensure that women are significantly represented in all capacity building programs

Output 3: Collaboration between different stakeholders is improved

– Ensure the involvement of all stakeholders, including representatives of farmers, in all stages of the approach from design to execution.
– Create a mechanism for data integration using all involved stakeholders' data
– Develop a software interface to allow data mapping and visualization
– Develop a recommendation system to offer decision guiding for both famers and policy makers

4 Conclusion

In this paper, a Theory of Change (ToC) approach to building a climate-resilient agriculture system in Morocco was presented. The approach will leverage the benefits of powerful Big Data Analytics to discover hidden patterns in the data and establish recommendation systems guiding both farmers and decision makers' choices.

Future work will consist in applying the proposed approach for a small-scale project targeting three agricultural regions in Morocco. Creation and integration of datasets will be based on real data derived from many stakeholders: Ministry of Agriculture, Direction of Strategy and Statistics, Ministry of Transportation, communities, and farmers.

References

1. Perry, M.: Moroccan agriculture: facing the challenges of a divided system (2015). http://sustainablefoodtrust.org/articles/moroccan-agriculture-facing-challenges-divided-system/
2. CGMS-MAROC: Comprehensive and Advanced Crop Monitoring System for Morocco. http://www.cgms-maroc.ma
3. ACCAGRIMAG. http://www.ffem.fr/accueil/projets/projets_ffem-par-secteur/Projetschangement-climatique/2013-CZZ1812-projet-FFEM-ACCAGRIMAG-Maroc-Tunisie
4. IRRISAT: Irrigation assisted by Satellite. http://www.irrisat.it
5. IRRiEYE. http://www.irrieye.com
6. Adekunle, A.A., Fatunbi, A.O.: A new theory of change in African agriculture. Middle-East J. Sci. Res. **21**(7), 1083–1096 (2014)
7. Mousannif, H., Sabah, H., Douiji, Y., Oulad Sayad, Y.: Big data projects: just jump right in! Int. J. Pervasive Comput. Commun. **12**(2), 260–288 (2016)
8. Mousannif, H., Sabah, H., Douiji, Y., Oulad Sayad, Y.: From big data to big projects: a step-by-step roadmap. In: Proceeding of International Conference of Future Internet of Things and Cloud (FiCloud). IEEE Xplore (2014)
9. Karkouch, A., Mousannif, H., Al Moatassime, H., Noel, T.: Data quality in internet of things: a state-of-the-art survey. J. Netw. Comput. Appl. **73**, 57–81 (2016)
10. Batini, C., Scannapieco, M.: Data quality: concepts, methodologies and techniques (2006). http://www.ciando.com/ebook/bid-35856-data-quality-concepts-methodologies-and-techniques/inhalte/
11. Wang, R., Strong, D.M.: Beyond accuracy: what data quality means to data consumers. J. Manage. Inf. Syst. **12**(4), 5 (1996)
12. Karkouch, A., Mousannif, H., Al Moatassime, H., Noel, T.: A model-driven framework for data quality management in the Internet of Things. J. Ambient Intell. Humanized Comput., 1–22 (2017)
13. Aggarwal, C.C., Ashish, N., Sheth, A.: The internet of things: a survey from the data-centric. In: Managing and Mining Sensor Data, pp. 383–428 (2013). Chapter 12
14. Soldatos, J., et al.: OpenIoT: open source Internet-of-Things in the cloud. In: Interoperability and Open-Source Solutions for the Internet of Things, pp. 13–25. Springer (2015)
15. Javed, N., Wolf, T.: Automated sensor verification using outlier detection in the internet of things. In: Proceedings - 32nd IEEE International Conference on Distributed Computing Systems Workshops, ICDCSW 2012, pp. 291–296 (2012)

16. Hofstra, N., Haylock, M., New, M., Jones, P., Frei, C.: Comparison of six methods for the interpolation of daily, European Climate Data. J. Geophys. Res. D: Atmos. **113**(21), D21110 (2008)
17. Štěpánek, P., Zahradníček, P., Huth, R.: Interpolation techniques used for data quality control and calculation of technical series: an example of a central european daily time series. Idojaras **115**(1–2), 87–98 (2011)
18. Lei, J., Bei, H., Xia, Y., Huang, J., Bae, H.: An in-network data cleaning approach for wireless sensor networks. Intell. Autom. Soft Comput. **8587**(March), 1–6 (2016)
19. Shashank, S., Wolf, T.: Massively parallel anomaly detection in online network measurement. In: Proceedings - International Conference on Computer Communications and Networks, ICCCN, vol. 1, pp. 261–266 (2008)
20. Thanigaivelan, N.K., Kanth, R.K., Virtanen, S., Isoaho, J.: Distributed internal anomaly detection system for internet-of-things. In: 2016 13th IEEE Annual Consumer Communications & Networking Conference (CCNC), pp. 0–1 (2016)
21. Mandagere, N., Zhou, P., Smith, M., Uttamchandani, S.: Demystifying data deduplication. In: Proceedings of the ACMIFIPUSENIX International Middleware Conference Companion on Middleware 08 Companion, vol. 08, pp. 12–17 (2008)

A Method for Segmentation of Agricultural Fields on Aerial Images with Markov Random Field Model

Jamal Bouchti[1](✉), Adel Asselman[1], and Abdellah El Hajjaji[2]

[1] Optique and Photonic Team, Faculty of Sciences, M'hannech II, Tetuan, Morocco
jamalbouchti@gmail.com

[2] Systems of Communications and Detection Laboratory, Faculty of Sciences, M'hannech II, Tetuan, Morocco

Abstract. Aerial imaging has become important to areas like remote sensing, surveying, and particularly in the agricultural application areas. In this paper, we suggest an aerial image segmentation approach based on Markov random field model and Gibbs distributions, we introduce iterative algorithm process to minimize an energy function which incorporate a local characteristics of pixel like color and also Neighborhood characteristics like texture and CIEDE2000.

Keywords: Aerial image segmentation · Markov random field · Gibbs distribution · Potential energy function · Iterative algorithm · CIEDE2000

1 Introduction

Image segmentation has been proved to be powerful tool for analysis in numerous fields and applications. The most advantages of using imaging technology for sensing are that it can be pretty accurate, nondestructive, and yields consistent results.

Applications of image processing technology will improve industry's productivity, thereby reducing costs and making agricultural operations and processing more secure for farmers and processing-line employees [1, 2].

Our method consists of the following steps:

First, we transform our image to the HSV color space which provides three components (Hue, Saturation, Value) interesting information, in particular to develop robust methods to changes in illumination. It is therefore possible to reduce the sensitivity to changes in illumination while remaining sensitive to lighting conditions and noise in shaded areas, taking into account only the components of Chrominance (hue and saturation).

Secondly, a method is introduced for labeling each pixel on the basis of its local characteristics such as color, texture or its proximity to a contour. We get a preliminary segmentation of our image.

The third step consists of constructing a model of Markov fields with conditional potential functions based on the weighting characteristics of the image segment. Then an iterative algorithm is computed which calculates the energy of each pixel with

M. Ezziyyani (Ed.): AI2SD 2018, AISC 911, pp. 98–105, 2019.
https://doi.org/10.1007/978-3-030-11878-5_11

respect to a given neighborhood system, and applying the principle of minimum potential energy to estimate a new segmentation of the image. Finally, we keep the best configuration of the image that corresponds to the final segmentation.

The reminder of this paper is organized as follows. In Sects. 2.1 and 2.2, we describe the HSV color space and texture features extraction which will be utilized to extract the list of homogeneous regions existing in the image. Section 2.3 gives a description of The CIE 2000 color difference formula introduced in the MRF model described in Sect. 2.4. In Sect. 3, we provide some results of the image segmentation method. Finally, the last section gives conclusions and future work.

2 Experimental and Computational Details

2.1 Generate Color Features

After loading image, we applied median filtering proposed by [3, 4]. This filter removes the image noise from the homogeneous regions, but preserves the edges in other areas. consequently, the image is transformed from RGB space to the HSV.

The most advantages of HSV are the facility to separate color component (hue) from intensity and saturation, the segmentation result is more robust and invariant to lighting changes.

Hue is described as an angle in the interval $[0, 2\pi]$ relative to the red axis, with red at angle 0 and 2π, green at $2\pi/3$ and blue at $4\pi/3$. Saturation de notes the perceived intensity of a specific color is measured as a radial distance from the central axis with value between 0 at the center to 1 at the outer surface. Value denotes brightness perception of a specific color. Then again, for a given Intensity and Hue, if the Saturation is changed from 0 to 1, the perceived color changes from a shade of gray to the purest form of the color represented by its Hue. For low values of Saturation, a color could be represented by a gray value specified by the Intensity level while for higher Saturation, the color could be represented by its Hue. The Saturation threshold that determines this transition once again depends on the Intensity [5].

Assuming the maximum Intensity value to be 255, we use the next threshold expression (1) to decide if a pixel must be represented by its Hue or its Intensity as its dominant feature [6].

$$\mathrm{th_{sat}}(\mathrm{V}) = 1.0 - \frac{0.8\,\mathrm{V}}{255} \tag{1}$$

We generate features by utilizing the above properties of the HSV color space for clustering pixels into segmented areas.

2.2 Generating Texture Features

Matrices of co-occurrences, proposed by Haralick in 1973 [7], constitute a powerful tool to study the distribution of color information, in particular in the case of chromatic co-occurrence matrices, while Taking into account the spatial relationship between the pixels.

Texture features, based on GLCM, are efficient means to study the texture of an image. Given the image composed of pixels each with an intensity, the GLCM is an illustration of how frequently different combinations of grey levels concur in that image [8].

To define an image, let us consider a finite lattice S with a neighborhood structure $\Re = \{Ns,\ s \in S\}$, where Ns is a neighborhood of the site s. Let I be a color HSV image defined on S, each site $s \in S$ is characterized by its color $I(s) = (I_h(s), I_s(s), I_v(s))$.

For each color component $k \in \{H,\ S,\ V\}$, we call color level set $S_g^k \subseteq S$ the set of sites with a given level $g \in [0\ldots q-1]$ in the color component image I_k, namely $S_g^k = \{s \in S,\ I_k(s) = g\}$. Then, $\left\{S_g^k\right\}$ is a partition of S for k, such that $\bigcup_{g=0}^{q-1} S_g^k = S$ and $S_g^k \cap S_{g'}^k = \varnothing$ for any pair of levels $g' \neq g'$. CLCMs that analyze each color component image independently are related to Chromatic CMs that characterize each pair of color component images [9]. Let $m_k[I]$ be the CLCM that captures the spatial interactions between the site levels I color component image I_k according to the neighborhood structure $\Re$. The cell $m_k[I]\ (g,\ g'),\ (g,\ g') \in [0\ldots q-1]2$ of this matrix contains the number of times that a site whose level is g' occurs in the neighborhood Ns of a site s whose level is g:

$$m^k[I](g, g') = \sum_{s \in S} \sum_{r \in Ns} \begin{cases} 1 & \text{if}\quad s \in S_g^k \text{ and } r \in S_{g'}^k \\ 0 & \text{otherwise} \end{cases} \tag{2}$$

The co-occurrence matrices contain a very large mass of information and are therefore difficult to manipulate, thus fourteen indices (Defined by Haralick) which correspond to Descriptive characters of textures can be calculated from these matrices. In this paper, we present and use only two of these indices:

Homogeneity. Homogeneity indicates the similarity of pixels. A diagonal value co-occurrence matrix provide homogeneity of 1. It becomes considerable if local textures only have minimal changes. It can be calculated as:

$$\text{Homogeneity} = \sum_{i=1}^{q-1} \sum_{j=1}^{q-1} \frac{P(i,j)}{1 + |i-j|} \tag{3}$$

Correlation. This attribute measures how correlated a pixel is to its neighborhood. It is the measure of value linear dependencies in the image. Feature values range from –1 to 1, these extremes indicating perfect negative and positive correlation respectively. μ_i and μ_j are the means and σ_i and σ_j are the standard deviations of P(i) and P(j), respectively. If the image has horizontal textures the correlation in the direction of 0° is often larger than those in other directions. It can be calculated as:

$$\text{Correlation} = \sum_{i=1}^{q-1} \sum_{j=1}^{q-1} \frac{ijP(i,j) - u_i u_j}{\sigma_i \sigma_j} \tag{4}$$

2.3 The CIEDE2000

The L*a*b* is a color space defined by the International Commission on Illumination (CIE) in 1976, consists of a lightness L* value and two color components, a* axis extends from green (–a) to red (+a) and the b* axis from blue (–b) to yellow (+b) (Fig. 1).

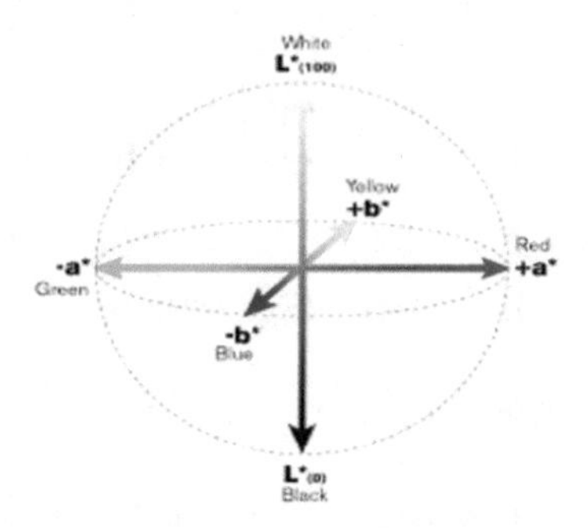

Fig. 1. L*a*b* color space

Members of CIE Technical Committee 1-47 develop the CIEDE2000 formula and was published by the CIE in 2001. The CIEDE2000 color difference equation provides an improved procedure for the computation of color differences between simple patches.

The ΔE_{00} equation is notably more sophisticated and computationally involved than its predecessor color-difference equations $\Delta E^{*}ab$ for CIELAB and the CIE94 color difference ΔE_{94} [10].

Given a couple of color values in CIELAB space L_1^*, a_1^*, b_1^* and L_2^*, a_2^*, b_2^*, and parametric weighting factors k_L, k_C, and k_H, the process of computation of the color difference ΔE_{00} is summarized in the following equations, grouped as three fundamental steps [11]:

Calculate C′i, h′i

$$C_{i,ab}^{*} = \sqrt{(a_i^*)^2 + (b_i^*)^2} \quad i = 1,2 \tag{5}$$

$$\overline{C}_{ab}^{*} = \frac{C_{1,ab}^{*} + C_{2,ab}^{*}}{2} \tag{6}$$

$$G = 0.5\left(1 - \sqrt{\frac{\overline{C}_{ab}^{*7}}{\overline{C}_{ab}^{*7} + 25^7}}\right) \tag{7}$$

$$a_i' = (1 + G)a_i^* \quad i = 1,2 \tag{8}$$

$$C_i' = \sqrt{(a_i^*)^2 + (b_i^*)^2} \quad i = 1,2 \tag{9}$$

$$h'_i = \begin{cases} 0 & b^*_i = a'_i = 0 \\ \tan^{-1}(b^*_i, a'_i) & \text{otherwise} \end{cases} \quad i = 1, 2 \tag{10}$$

Calculate $\Delta L', \Delta C', \Delta H'$

$$\Delta L' = L^*_2 - L^*_1 \qquad \Delta C' = C'_2 - C'_1 \tag{11}$$

$$\Delta L' = \begin{cases} 0 & C'_1 C'_2 = 0 \\ h'_2 - h'_1 & C'_1 C'_2 \neq 0; \left|h'_2 - h'_1\right| \leq 180^\circ \\ (h'_2 - h'_1) - 360 & C'_1 C'_2 \neq 0; \left(h'_2 - h'_1\right) > 180^\circ \\ (h'_2 - h'_1) + 360 & C'_1 C'_2 \neq 0; \left(h'_2 - h'_1\right) < -180^\circ \end{cases} \tag{12}$$

Calculate ΔE00 Color-Difference

$$\bar{L}' = (L^*_1 + L^*_2)/2 \qquad \bar{C}' = (C'_1 + C'_2)/2 \tag{13}$$

$$\bar{h}' = \begin{cases} \frac{h'_1 + h'_2}{2} & \left|h'_1 - h'_2\right| \leq 180^\circ; C'_1 C'_2 \neq 0 \\ \frac{h'_1 + h'_2 + 360^\circ}{2} & \left|h'_1 - h'_2\right| > 180^\circ; (h'_1 + h'_2) < 360^\circ; C'_1 C'_2 \neq 0 \\ \frac{h'_1 + h'_2 - 360^\circ}{2} & \left|h'_1 - h'_2\right| > 180^\circ; (h'_1 + h'_2) \geq 360^\circ; C'_1 C'_2 \neq 0 \\ (h'_1 + h'_2) & C'_1 C'_2 = 0 \end{cases} \tag{14}$$

$$T = 1 - 0.17\cos\left(\bar{h}' - 30^\circ\right) + 2.24\cos\left(2\bar{h}'\right) + 0.32\cos\left(3\bar{h}' + 6^\circ\right) - 0.20\cos\left(4\bar{h}' - 63^\circ\right) \tag{15}$$

$$\Delta\theta = 30\exp\left\{-\left[\frac{\bar{h}' - 275^\circ}{25}\right]^2\right\} \tag{16}$$

$$R_c = 2\sqrt{\frac{\bar{C}'^7}{\bar{C}'^7 + 25^7}} \qquad S_L = 1 + \frac{0.015\left(\bar{L}' - 50\right)^2}{\sqrt{20 + \left(\bar{L}' - 50\right)^2}} \tag{17}$$

$$S_C = 1 + 0.045\bar{C}' \quad S_H = 1 + 0.015\bar{C}'T \quad R_T = -\sin(2\Delta\theta)R_C \tag{18}$$

$$\Delta E^{12}_{00} = \Delta E_{00}(L^*_1, a^*_1, b^*_1, L^*_2, a^*_2, b^*_2)$$

$$\Delta E^{12}_{00} = \sqrt{\left(\frac{\Delta L'}{K_L S_L}\right)^2 + \left(\frac{\Delta C'}{K_C S_C}\right)^2 + \left(\frac{\Delta H'}{K_H S_H}\right)^2 + R_T\left(\frac{\Delta C'}{K_C S_C}\right)\left(\frac{\Delta H'}{K_H S_H}\right)} \tag{19}$$

2.4 Markov Random Field

A Markov Random Field (MRF) is a graphical model of a joint probability distribution. It consists of an undirected graph G = (N, Ɛ) in which the nodes N represent random variables [12].

Let S be a set of locations (pixels) S = {(i, j) | i, j are integers}. Neighbors of S are defined as:

$$N(i,\ j) = \left\{(k,\ l)\,|\,0 < (k - i)^2 + (l - j)^2 < \text{constant}\right\} \tag{20}$$

A subset C of S is a clique if any two different elements of C are neighbors. The set of all cliques of S is denoted by Ω (Fig. 2).

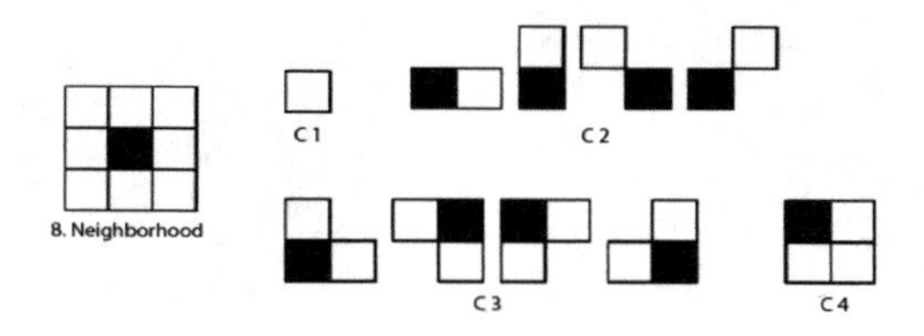

Fig. 2. Cliques for 8 neighborhoods

The random vector on S is called a random field and assumed to have density P(x).

$$P(X_s = x_s|X_t = x_t, \forall t \neq s) = P(X_s = x_s|X_t = x_t, \forall t \in N\{s\}) \tag{21}$$

The value Xs at location S depends only on its neighbors. P(x) can also be factorized over cliques due to its Markov properties.

Hamersley Clifford Theorem defines any positive distribution that obeys the Markov property equivalent to Gibbs Fields [13] and P(x) can be written in the form:

$$P(x) = \frac{1}{Z}\exp(-U(x)) \tag{22}$$

The denominator Z is a normalizing constant named the partition function, U(x) is the energy function, and is the sum of clique potential over all possible cliques C.

After introducing an image labeling field which is an unobservable MRF, the segmentation problem turns into an optimization problem. In the optimization process, we use the posterior probability value to label pixels to different classes.

$$X = argmin\ U(x) \tag{23}$$

3 Result

Following are a sample results obtained by going through the various stages of our method, we can perceive that there are very small regions that we can eliminate by introducing a fusion condition to our algorithm, another improvement we are studying to use multithreading [14, 15] programming in our algorithm since the computation of the probability region of a pixel depends only on the state of its neighbors (Fig. 3).

Fig. 3. (Left) the Original image, (right) the Final region segmentation result.

4 Conclusion

In this paper, we presented a probabilistic method to segment aerial agriculture image. In the first step, image is stocked in the memory and converted into HSV color space. Then, the pre-processing procedure is applied to remove noise caused by the conversion process. Moreover, the HSV color feature extracted from the image. Regarding the extraction of the texture feature, the Color Co-occurrence Matrix (CCM) is used, which includes Correlation and Homogeneity. In the final stage, on the basis of the results we obtained, the image is segmented by using an MRF and iterative algorithm including the CIEDE2000 formula in the energy function.

References

1. Nasir, F.A., et al.: A study of image processing in agriculture application under high performance computing environment (2012)
2. Janwale, A.: Digital image processing applications in agriculture: a survey. Int. J. Adv. Res. Comput. Sci. Softw. Eng. **5**, 622 (2015)
3. Verma, K., Singh, B.K., Thokec, A.S.: An enhancement in adaptive median filter for edge preservation. In: International Conference on Intelligent Computing, Communication & Convergence (ICCC-2015) (2015)
4. Carron, T., Lambert, P.: Color edge detector using jointly hue, saturation and intensity. In: Proceedings of the IEEE International Conference on Image Processing, ICIP 1994, vol. 3, pp. 977–981 (1994)

5. El Asnaoui, K., Aksasse, B., Ouanan, M.: Content-based color image retrieval based on the 2-D histogram and statistical moments. In: Second World Conference on Complex Systems (WCCS), Agadir, pp. 653–656 (2014)
6. Sural, S., Qian, G., Pramanik, S.: Segmentation and histogram generation using the HSV color space for image retrieval. In: Proceedings of the International Conference on Image Processing, vol. 2, p. II (2002)
7. Haralick, R.M., Shanmugam, K., Dinstein, I.: Textural features for image classification. IEEE Trans. Syst. Man Cybern. **3**, 610–621 (1973). https://doi.org/10.1109/TSMC.1973.4309314
8. Bayram, U., Can, G., Duzgun, S., Yalabik, N.: Evaluation of textural features for multispectral images, p. 81800I (2011)
9. Guerrout, E.-H., Mahiou, R., Ait-Aoudia, S.: Medical image segmentation on a cluster of PCs using Markov random fields. Int. J. New Comput. Archit. Appl. (IJNCAA) **3**(1), 35–44 (2013)
10. Gómez-Polo, C., Muñoz, M.P., Luengo, M.C.L., Vicente, P., Galindo, P., Casado, A.M.M.: Comparison of the CIELab and CIEDE2000 color difference formulas. J. Prosthet. Dent. **115** (1), 65–70 (2016). https://doi.org/10.1016/j.prosdent.2015.07.001. Published online 26 Sep 2015
11. Sharma, G., Wu, W., Dalal, E.N.: The CIEDE2000 color-difference formula: implementation notes, supplementary test data, and mathematical observations. Color Res. Appl. **30**, 21–30 (2005)
12. Geman, S., Geman, D.: Stochastic relaxation, Gibbs distributions, and the Bayesian restoration of images. IEEE Trans. Pattern Anal. Mach. Intell. **6**, 721–741 (1984)
13. Cruz, H., Eckert, M., Meneses, J.M., Martínez, J.F.: Fast evaluation of segmentation quality with parallel computing. Sci. Program. **2017**, 9 (2017). Article ID 5767521
14. Dey, N., Mukherjee, A., Madhulika, Chakraborty, S., Samanta, S.: Parallel image segmentation using multi-threading and k-means algorithm. In: IEEE International Conference on Computational Intelligence and Computing Research, IEEE ICCIC 2013 (2013). https://doi.org/10.1109/iccic.2013.6724171
15. Sagheb, E.: A Survey of Multithreading Image Analysis (2015)

Fuzzy TOPSIS with Coherent Measure: Applied to a Closed Loop Agriculture Supply Chain

Mohamed El Alaoui[1](✉), Hussain Ben-Azza[1], and Khalid El Yassini[2]

[1] Department of Industrial and Production Engineering, ENSAM-Meknes, Moulay Ismail University, Meknes, Morocco
mohamedelalaoui208@gmail.com

[2] IA Laboratory, Science Faculty, Moulay Ismail University, Meknes, Morocco

Abstract. Fuzzy logic has been widely used combined with Multi Criteria Decision Making techniques in different application. Here we propose to aggregate fuzzy opinions with a mathematical model, in order to minimize discordances. An illustrative example treating closed loop agriculture Supply Chain is given.

Keywords: Fuzzy TOPSIS · Optimal weight · Coherence measure · Performance analysis · Agriculture

1 Introduction

Multi Criteria Decision Making (MCDM) methods referred also Multi Criteria Decision Analysis (MCDA) [1] or Multi-Attribute Decision Making (MADM) [2], have been used in several domains, Supply Chain Management [3], Logistics [4, 5], Health [6, 7], Safety and Environment [8, 9], Energy Management [10, 11], Water Resources Management [12] and many other topics. The Technique for Order Preference by Similarity to Ideal Solution (TOPSIS) happens to be a satisfactory MCDM technique that can fit in various areas [4–7].

To able to model the human reasoning, Zadeh [13] introduced fuzzy sets, that were adapted later to the decision making framework [14] by introducing the notion of a linguistic variable. In fact it is easier for a decision maker (DM) to qualify an alternative by a linguistic (good, bad, average …) than to give a crisp assessment. A natural junction was made after between fuzzy logic and TOPSIS [15].

Several MCDM methods, consider equal weighting for DMs according to all criteria. However, no DM can claim a perfect knowledge according to all criteria of selection. Here we propose a coherence based method to weight DM opinions' by adapting the algorithm proposed by Lee [16] aiming to minimize the weighted incoherencies between the expressed fuzzy opinions.

The rest of this manuscript is as follows: Sect. 2 contains the required definitions. Section 3 presents a review of MCDM methods and their relative applications in agriculture decision making. Section 4 introduced the needed background from fuzzy

M. Ezziyyani (Ed.): AI2SD 2018, AISC 911, pp. 106–117, 2019.
https://doi.org/10.1007/978-3-030-11878-5_12

logic to represent DM uncertainty. In Sect. 5, the algorithm promoting coherent opinions is presented. Section 6 describes the methodology proposed approach. Section 7 contains an illustrative example. Section 8 concludes this work.

2 Definitions

2.1 Supply Chain (SC)

There is no consensual definition of a SC, since each expert will define it according to his position on the SC. However, all of them will agree that a supply chain contains suppliers, manufacturers, distributors, retailers and final users.

2.2 Supply Chain Management (SCM)

The Council of Logistics Management (CLM) [17] defines SCM as: "*the systemic, strategic coordination of the traditional business functions and tactics across these businesses functions within a particular organization and across businesses within the supply chain for the purposes of improving the long-term performance of the individual organizations and the supply chain as a whole*".

2.3 Agriculture Supply Chain Management (ASCM)

The negative influence of the human activity on our environment, imposes considering the notion of Green supply chain management (GSCM) while designing new SCs. The GSCM aims minimizing negative SC impact on the environment involving the whole SC, starting from product design to the final user and the feedback [18]. Since agriculture, afford raw materials for many SCs, the notion of ASCM imposes itself as major component of GSCM, as well as a sell label.

The ASCM implies monitoring the relationships from the farm level to the final consumers, to meet consumers' requirements in terms of quantity, quality and price.

3 MCDM and Agriculture

Fast technological advancement caused premature disposal of products and shortened product life cycle, which raises the environmental awareness about long term impacts, MCDM techniques have been widely used to model such conflicting situations [19]. But first why multi and not mono criteria? According to Zeleney [20], there is no decision making with less than two criteria. In fact, when maximizing the land productivity, we are in most of cases using some pesticides that have negative impacts on environmental issues. This example, illustrate that real life situations that we want to see as mono-criteria, turns easily in reality into a multi-criteria problems.

Tarrant [21] divided decision making in agriculture into three categories. In the first, the physical environment imposes the decisions that can be considered. In the second economy would be dictating the rules. While in the third, personal values, risk aversion can have an enormous impact in the final decision.

Joining the last category in Tarrant's study, Gasson [22] asked 100 farmers to rank 60 attributes according to their working values. The author raised since that the conflicting nature of the studied indicators, such as economy, expansion and independence.

Yemshanov et al. [23] evaluate land clearing potential in the Canadian agriculture forestry. They compared the expected land value if it remains forested, and if it is cleared for agriculture. They aggregate several indicators into a single one representing the overall satisfaction.

Al-Juaid et al. [24], proposed a methodology for water allocation in deficit region, Gaza, Palestine, considering economic issues, environmental aspects, and public health risks.

Chen et al. [25] examined land suitability in order to expand or retire retrieve irrigation, the use of fuzzy linguistic in their approach, permits incorporating uncertainty. other approaches to evaluate land suitability can be found in [26–29].

Sinha and Tripathi [30] proposed using MCDM involving agro-meteorological data to improve crop insurance. They claim that their method can afford reasonable prediction with less historical data. Especially in emerging countries.

Petkovics et al. [31] attempted to optimize drone choice for precision agriculture. They argued that drones may best feet than intelligent sensors in collecting the heterogeneous required data for crops.

Romero and Rehman [32] treated the essence of MCDM and the logical structure for several agriculture problems with an emphasis in the mathematical programming background. They also provided various applications and pointed out the role of risk and uncertainty in the agriculture decision making.

Further application of MCDM methods in the agriculture framework can be found in [33–36].

4 Fuzzy Logic

Several authors stressed on the importance of considering uncertainty in agriculture decision making [25, 26, 32, 37–40]. Thus, a natural question can be formulated is: what is uncertainty? We must agree first that there is no unanimous consensus about its definition. Klir and Folger [41] quoted six different definitions. A comprehensive review of uncertainty models in MCDM was proposed in [42].

To deal quantitatively with the ambiguity of human judgement, L. Zadeh introduced fuzzy sets [13]. Let X be a universal set and F a fuzzy subset in X, F is defined as follows [43]:

$$F = \{ <x, \mu_F(x)> \, | x \in X \}$$

where $\mu_F(x)$ is the degree of membership of x in F in the unity interval.

$$\mu_F : X \rightarrow [0, 1].$$

A fuzzy set is convex if and only if:

$$\mu_F(\lambda x_1 + (1-\lambda)x_2) \geq \min(\mu_F(x_1), \mu_F(x_2))$$

for all $x_1, x_2 \in X$ and $\lambda \in [0,1]$.

A fuzzy set F is normalized if $\sup(\mu_F) = 1$.

A fuzzy number is a convex normalized fuzzy set of the real line R^1 whose membership function is piecewise continuous.

A fuzzy number N is called positive, denoted by $N > 0$, if its membership function $\mu_N(x) = 0$ for all $x < 0$.

To ease computation, we will adopt triangular fuzzy numbers as used in many other researchers [3, 15].

A positive triangular fuzzy number $\tilde{A}$ is denoted by 3-tuples (a_1, a_2, a_3), as follows:

$$\mu_{\tilde{A}}(x) = \begin{cases} \frac{x-a_1}{a_2-a_1}, & a_1 \leq x \leq a_2 \\ \frac{x-a_3}{a_2-a_3}, & a_2 \leq x \leq a_3 \\ 0, & otherwise \end{cases}$$

For any two positive triangular fuzzy numbers $\tilde{A}(a_1, a_2, a_3)$ and $\tilde{B}(b_1, b_2, b_3)$ the fuzzy addition, subtraction, multiplication, inverse and division are respectively defined as follows:

$$\tilde{A} \oplus \tilde{B} = (a_1 + b_1, a_2 + b_2, a_3 + b_3)$$
$$\tilde{A} \ominus \tilde{B} = (a_1 - b_3, a_2 - b_2, a_3 - b_1)$$
$$\tilde{A} \otimes \tilde{B} = (a_1 * b_1, a_2 * b_2, a_3 * b_3)$$
$$1 \oslash \tilde{B} = (1/b_3, 1/b_2, 1/b_1)$$
$$\tilde{A} \oslash \tilde{B} = (a_1/b_3, a_2/b_2, a_3/b_1)$$

The distance between two triangular fuzzy numbers $\tilde{A}(a_1, a_2, a_3)$ and $\tilde{B}(b_1, b_2, b_3)$ can be computed as follows:

$$D(\tilde{A}, \tilde{B}) = \sqrt{\frac{1}{3}\left[(a_1 - b_1)^2 + (a_2 - b_2)^2 + (a_3 - b_3)^2\right]}$$

5 Coherence Algorithm

Weights can influence mainly the final result. Al-Kloub et al. [44] studied weights sensitivity to select water project in Jordan. A comprehensive review of weighting methods and their effects on MCDM outcomes in water resource management was proposed in [45].

Here we adopt the algorithm proposed by Lee [16], in order to compute the optimal weights. The optimized function can be formulated as follows:

$$\min_{M\times IR^3} \sum_{k=1}^{K} w_k^q * (c - S(\tilde{x}_k, \tilde{x})),$$

where $M = \left\{ \begin{array}{c} W = (w_1, w_2, \ldots, w_K), w_k \geq 0, \\ \sum_{j=1}^{m} w_j = 1, \end{array} \right\}$

m is a positive integer $q > 1$, c is a real number $c > 1$, $\tilde{x}_k(x_{k,1}, x_{k,2}, x_{k,3})$ are the individual opinions expressed by each DM. $\tilde{x}(x_1, x_2, x_3)$ is the aggregated consensus $S(\tilde{x}_k, \tilde{x})$ the similarity between the kth decision and the consensus.

The similarity can be adapted to triangular fuzzy numbers as follows:

$$S(\tilde{x}_k, \tilde{x}) = 1 - \frac{1}{4u^2}(D(\tilde{x}_k, \tilde{x}))^2$$

where $u = \max_{k,t}(x_{j,t}) - \min_{k,t}(x_{k,t})$ and $t = [|1, 2, 3|]$

This problem can be resolved by the following algorithm:

Algorithm 1:

Step 1: each DMs express his opinion by a triangular fuzzy number $\tilde{x}_k(x_{k,1}, x_{k,2}, x_{k,3})$

Step 2: fix the initial weights $W^{(0)}\left(w_1^{(0)}, \ldots, w_K^{(0)}\right)$ verifying $0 \leq w_k^{(0)} \leq 1$ and $\sum_{k=1}^{K} w_k^{(0)} = 1$. The iterations will be marked by $l = 0, 1, \ldots$

Step 3: compute

$$\tilde{x}^{(l+1)} = \frac{\sum_{k=1}^{K} w_k^{(l)^q} \otimes \tilde{x}_k}{\sum_{k=1}^{K} w_k^{(l)^q}}$$

Step 4: compute

$$w_j^{(l+1)} = \frac{\left(1/\left(c - S\left(\tilde{x}_k, \tilde{x}^{(l+1)}\right)\right)\right)^{1/(q-1)}}{\sum_{j=1}^{m}\left(1/\left(c - S\left(\tilde{x}_k, \tilde{x}^{(l+1)}\right)\right)\right)^{1/(q-1)}}$$

Step 5: if $\left\|W^{(l+1)} - W^{(l)}\right\| \leq \varepsilon$ stop, else set $l = l + 1$ and go to step 3.

6 Proposed Approach

TOPSIS was proposed by Hwang and Yoon in 1981 [46], it aims to choose the alternative which is in the closest to the positive ideal solution (PIS) and the furthest from the negative ideal solution (NIS), this can be computed by the closeness coefficient. But upon all other methods, what makes TOPSIS more relevant? First TOPSIS can deal with subjective and objective criteria. Suppose an MCDM situation involving new pesticides that are not well known, economic results and farmers working values. In addition TOPSIS requires little input from the DM [47] compared to other frequently used methods.

The problem is formulated as follows:

K DMs $k = (1, \ldots, K)$, evaluate n alternatives $A = (A_1, A_2, \ldots, A_n)$, according to m criteria $C = (C_1, C_2, \ldots, C_m)$.

Each DM will assess criteria weight $\tilde{W} = \left(\tilde{w}_{jk}; j = 1, \ldots, m; k = 1, \ldots, K\right)$ and alternative in accordance to criteria of evaluation $X = \left(\tilde{x}_{ijk}; i = 1, \ldots, n; j = 1, \ldots, m; k = 1, \ldots, K\right)$.

The aggregation of individual assessment will occur according to the following algorithm 1, the rest will follow the classical TOPSIS algorithm.

The proposed method can be summarized as follows:

Step 1: each DM assesses criteria importance by linguistic variables (Table 1).

Step 2: each DM assesses alternatives with respect to each criterion (Table 1).

Step 3: convert the linguistic variables to triangular fuzzy numbers (Table 1).

Step 4: construct the collective assessment for each alternative in accordance to each criterion $\tilde{x}_{ij}$ and the collective assessment for criteria weights $\tilde{w}_j$ using Algorithm 1.

Step 5: compute the weighted decision for each decision

$$\tilde{G}_{ij} = \tilde{x}_{ij} \otimes \tilde{w}_j; i = 1, \ldots, n; j = 1, \ldots, m$$

Step 6: determine the PIS $\tilde{S}_j^+ = (9, 10, 10)$ and the NIS $\tilde{S}_j^- = (0, 0, 1)$

Step 7: compute the distance to each ideal solution

$$D_i^- = \sum\nolimits_{j=1}^{m} D\left(\tilde{G}_{ij}, \tilde{S}_j^-\right) i = 1, \ldots, n$$

$$D_i^+ = \sum\nolimits_{j=1}^{m} D\left(\tilde{G}_{ij}, \tilde{S}_j^+\right) i = 1, \ldots, n$$

Step 8: compute the closeness coefficients

$$CC_i = \frac{D_i^-}{D_i^- + D_i^+}$$

where D_i^- is the distance between the alternative assessment and the NIS and D_i^+ the distance between the alternative assessment and the PIS.

Step 9: rank alternatives according to the closeness coefficients

Since all evaluation occurs in the same scale, no normalization is required. We mention also that the proposed approach compared to the initial fuzzy TOPSIS [15], results in less uncertain result. Let $H(\tilde{x}_k)$ the uncertainty, also called hesitancy, for the kth opinion computed as follows:

$$H(\tilde{x}_k) = \int_{-\infty}^{+\infty} \mu_{\tilde{x}_k}(x)dx$$

Suppose three DMs asses an alternative by (VL, VL, VH) (Table 1) the aggregated result according to [15] is: (0,11/3,10) hence the hesitancy worth 5, which represents the half of all possible uncertainty. While the aggregated result in the proposed is (1.2685, 2.4497, 4.2685), leading to a significantly lower hesitancy equal to 1.5.

Table 1. Criteria importance/alternative evaluation.

Linguistic variable	Fuzzy number
Extremely low EL	$(0, 0, 0.1)$
Very low VL	$(0, 0.1, 0.3)$
Low L	$(0.1, 0.3, 0.5)$
Medium low ML	$(0.2, 0.4, 0.6)$
Medium M	$(0.3, 0.5, 0.7)$
Medium high MH	$(0.4, 0.6, 0.8)$
High H	$(0.5, 0.7, 0.9)$
Very high VH	$(0.7, 0.9, 1)$
Extremely high EH	$(0.9, 1, 1)$

We can easily prove that the proposed fulfills the following properties:

If $\tilde{x}_{k'} = \tilde{x}_{k''}$ for all k', k'' then $\tilde{x}_k = \tilde{x}$.

The uncertainty of aggregated decision for each alternative is between the uncertainties of all DMs relatively to the same alternative: $\min_k H(\tilde{x}_k) \leq H(\tilde{x}) \leq \max_k H(\tilde{x}_k)$.

The aggregated result is order independent: $\tilde{x} = f(\tilde{x}_1, \ldots, \tilde{x}_K) = f\left(\tilde{x}_{\sigma(1)}, \ldots, \tilde{x}_{\sigma(K)}\right)$, where σ is a permutation.

The common intersection of all DMs opinions is included in the aggregated consensus: $\bigcap_{k=1}^{K} \tilde{x}_k \subset \tilde{x}$.

7 Illustrative Example

Several approaches in the literature used TOPSIS in agriculture decision making [34, 48–53]. Seyedmohammadi et al. insisted on the fuzzy version of TOPSIS [54] for cultivation priority planning. In the sequel, we treat the example proposed by Tan et al. [55].

Three closed loop ASCM projects are evaluated by three DMs in accordance to four criteria (reliability, effectiveness cost, and asset management). Table 2 resumes the individual evaluations.

Table 2. Linguistic assessments.

DM	Reliability			Effectiveness			Cost			Assess management		
	A1	A2	A3	A1	A2	A3	A1	A2	A3	A1	A2	A3
DM1	H	M	H	EH	VH	VH	M	H	EH	M	H	EH
DM2	M	M	H	VH	M	H	M	M	VH	M	H	H
DM3	H	H	VH	EH	VH	EH	H	VH	EH	H	VH	EH

Table 3 resumes the iterative scheme for the first alternative (A1), according to the first criteria (Reliability).

Table 3. Linguistic assessments.

Iterations	DMs weights			Aggregated evaluation		
	DM1	DM2	DM3	A1	A2	A3
0	0.35000000	0.30000000	0.35000000	0.44626866	0.64626866	0.84626866
1	0.34139047	0.31721907	0.34139047	0.43969371	0.63969371	0.83969371
2	0.34027280	0.31945439	0.34027280	0.43882236	0.63882236	0.83882236
3	0.34012385	0.31975230	0.34012385	0.43870595	0.63870595	0.83870595
4	0.34010393	0.31979213	0.34010393	0.43869038	0.63869038	0.83869038
5	0.34010127	0.31979746	0.34010127	0.43868830	0.63868830	0.83868830
6	0.34010092	0.31979817	0.34010092	0.43868802	0.63868802	0.83868802
7	0.35000000	0.30000000	0.35000000	0.44626866	0.64626866	0.84626866

The final result remains the same independently from the starting point [16]. Table 4 resumes the collective decision for each alternative.

Table 4. Collective decision.

	A1	A2	A3
Di-	0.649504	0.638944	0.824076
Di+	0.308728	0.327389	0.127959
CCi	0.67781	0.66121	0.86559

It is true that the obtained ranking result, is similar to the one proposed in [55], making the 3rd alternative A3 largely superior to the others. However the distinction occurs while distinguishing the alternatives A1 and A2. TOPSIS consists on choosing the furthest solution from the NIS and the closest to the PIS. In [55], $D(A1, PIS) > D(A2, PIS)$ and $D(A1, NIS) > D(A2, NIS)$ leaving the final decision to the closeness coefficients. In the proposed, A1 is closer to the PIS than A2: $D(A1, PIS) < D(A2, PIS)$ and further from the NIS than A2: $D(A1, NIS) > D(A2, NIS)$

making clearly A1 in the second position and A2 in the third. This is due to the normalizing step and the interpretation of the PIS and the NIS. While same authors consider the absolute possible [15]. Others consider only the achieved solutions [55], which affect both the distances and the normalization.

As we mentioned before, the hesitancy of the aggregated opinions are much smaller.

8 Conclusion

TOPSIS is a user friendly method that require little input from the user, the method and its extended fuzzy version have already proved their efficiency and accuracy in several applications. Here we modified the aggregation procedure for the collective assessment, in order to minimize the output hesitancy and promote coherent opinions. It is true that the proposed modification is cumbersome in term of complexity, however the obtained result are more acceptable since achieving a better consensus. The comparative example applied to ASCM shows the effect of the classical normalizing procedure and the interpretation of ideal solutions on the final decision.

In future works, we may consider using type 2 fuzzy sets permitting a better modelling of the human uncertainty. An investigation of the algorithm parameters could be considered.

References

1. Ortuño, M.T.: Multi-criteria Decision Analysis (MCDA). In: Encyclopedia of Sciences and Religions, p. 1376. Springer, Dordrecht (2013)
2. Zavadskas, E.K., Turskis, Z., Kildienė, S.: State of art surveys of overviews on MCDM/MADM methods. Technol. Econ. Dev. Econ. **20**, 165–179 (2014)
3. Chen, C.-T., Lin, C.-T., Huang, S.-F.: A fuzzy approach for supplier evaluation and selection in supply chain management. Int. J. Prod. Econ. **102**, 289–301 (2006)
4. Chen, K.-H., Liao, C.-N., Wu, L.-C.: A selection model to logistic centers based on TOPSIS and MCGP methods: the case of airline industry. https://www.hindawi.com/journals/jam/2014/470128/
5. Hsueh, J.-T., Lin, C.-Y.: Integrating the AHP and TOPSIS decision processes for evaluating the optimal collection strategy in reverse logistic for the TPI. Int. J. Green Energy **14**, 1209–1220 (2017)
6. Mahdevari, S., Shahriar, K., Esfahanipour, A.: Human health and safety risks management in underground coal mines using fuzzy TOPSIS. Sci. Total Environ. **488–489**, 85–99 (2014)
7. Taylan, O., Zytoon, M.A., Morfeq, A., Al-Hmouz, R., Herrera-Viedma, E.: Workplace assessment by fuzzy decision tree and TOPSIS methodologies to manage the occupational safety and health performance. J. Intell. Fuzzy Syst. **33**, 1209–1224 (2017)
8. Kaya, B.Y., Dağdeviren, M.: Selecting occupational safety equipment by MCDM approach considering universal design principles. Hum. Factors Ergon. Manuf. Serv. Ind. **26**, 224–242 (2016)
9. Rezaian, S., Jozi, S.A.: Health-safety and environmental risk assessment of refineries using of multi criteria decision making method. APCBEE Procedia. **3**, 235–238 (2012)

10. Turgut, Z.K., Tolga, A.Ç.: Fuzzy MCDM methods in sustainable and renewable energy alternative selection: fuzzy VIKOR and fuzzy TODIM. In: Energy Management—Collective and Computational Intelligence with Theory and Applications, pp. 277–314. Springer, Cham (2018)
11. Büyüközkan, G., Karabulut, Y., Güler, M.: Strategic renewable energy source selection for turkey with hesitant fuzzy MCDM method. In: Energy Management—Collective and Computational Intelligence with Theory and Applications, pp. 229–250. Springer, Cham (2018)
12. Majumder, P., Saha, A.K.: Development of financial liability index for hydropower plant with MCDM and neuro-genetic models. In: Application of Geographical Information Systems and Soft Computation Techniques in Water and Water Based Renewable Energy Problems, pp. 71–105. Springer, Singapore (2018)
13. Zadeh, L.A.: Fuzzy sets. Inf. Control **8**, 338–353 (1965)
14. Bellman, R.E., Zadeh, L.A.: Decision-making in a fuzzy environment. Manag. Sci. **17**, B141–B164 (1970)
15. Chen, C.-T.: Extensions of the TOPSIS for group decision-making under fuzzy environment. Fuzzy Sets Syst. **114**, 1–9 (2000)
16. Lee, H.-S.: Optimal consensus of fuzzy opinions under group decision making environment. Fuzzy Sets Syst. **132**, 303–315 (2002)
17. USC of LM: What it's all about–purpose, objectives, programs, policies. The Council (1999)
18. Rao, P.: Greening of suppliers/in-bound logistics—in the South East Asian context. In: Sarkis, J. (ed.) Greening the Supply Chain, pp. 189–204. Springer, London (2006)
19. Gupta, S.M., Ilgin, M.A.: Multiple Criteria Decision Making Applications in Environmentally Conscious Manufacturing and Product Recovery. CRC Press, Boca Raton (2017)
20. Zeleny, M. (ed.): MCDM: Past Decade and Future Trends: A Source Book of Multiple Criteria Decision Making. JAI Press, Greenwich (1985)
21. Tarrant, J.R.: Tarrant: Agricultural Geography. Wiley, New York (1973)
22. Gasson, R.: Goals and values of farmers. J. Agric. Econ. **24**, 521–542 (1973)
23. Yemshanov, D., Koch, F.H., Riitters, K.H., McConkey, B., Huffman, T., Smith, S.: Assessing land clearing potential in the Canadian agriculture–forestry interface with a multi-attribute frontier approach. Ecol. Indic. **54**, 71–81 (2015)
24. Al-Juaidi, A.E., Kaluarachchi, J.J., Kim, U.: Multi-criteria decision analysis of treated wastewater use for agriculture in water deficit regions1. JAWRA J. Am. Water Resour. Assoc. 46, 395–411 (2010)
25. Chen, Y., Khan, S., Paydar, Z.: To retire or expand? A fuzzy GIS-based spatial multi-criteria evaluation framework for irrigated agriculture. Irrig. Drain. **59**, 174–188 (2010)
26. Benke, K.K., Pelizaro, C., Lowell, K.E.: Uncertainty in multi-criteria evaluation techniques when used for land suitability analysis. In: Crop Modeling and Decision Support, pp. 291–298. Springer, Heidelberg (2009)
27. Zolekar, R.B., Bhagat, V.S.: Multi-criteria land suitability analysis for agriculture in hilly zone: remote sensing and GIS approach. Comput. Electron. Agric. **118**, 300–321 (2015)
28. Ahmed, G.B., Shariff, A.R.M., Balasundram, S.K., bin Abdullah, A.F.: Agriculture land suitability analysis evaluation based multi criteria and GIS approach. IOP Conf. Ser. Earth Environ. Sci. **37**, 012044 (2016)
29. Aldababseh, A., Temimi, M., Maghelal, P., Branch, O., Wulfmeyer, V.: Multi-criteria evaluation of irrigated agriculture suitability to achieve food security in an arid environment. Sustainability **10**, 803 (2018)
30. Sinha, S., Tripathi, N.K.: Hybrid satellite agriculture drought indices: a multi criteria approach to improve crop insurance. In: 2016 Fifth International Conference on Agro-Geoinformatics (Agro-Geoinformatics), pp. 1–5 (2016)

31. Petkovics, I., Simon, J., Petkovics, Á., Čović, Z.: Selection of unmanned aerial vehicle for precision agriculture with multi-criteria decision making algorithm. In: 2017 IEEE 15th International Symposium on Intelligent Systems and Informatics (SISY), pp. 000151–000156 (2017)
32. Romero, C., Rehman, T.: Multiple Criteria Analysis for Agricultural, vol. 11, 2nd edn. Elsevier Science, Amsterdam, Boston (2003)
33. Bausch, J.C., Bojórquez-Tapia, L., Eakin, H.: Agro-environmental sustainability assessment using multicriteria decision analysis and system analysis. Sustain. Sci. **9**, 303–319 (2014)
34. Riesgo, L., Gallego-Ayala, J.: Multicriteria analysis of olive farms sustainability: an application of TOPSIS models. In: Handbook of Operations Research in Agriculture and the Agri-Food Industry, pp. 327–353. Springer, New York (2015)
35. López, J.C.L., Carrillo, P.A.Á., Valenzuela, O.A.: A multicriteria group decision model for ranking technology packages in agriculture. In: Soft Computing for Sustainability Science, pp. 137–161. Springer, Cham (2018)
36. Berbel, J., Bournaris, T., Manos, B., Matsatsinis, N., Viaggi, D. (eds.): Multicriteria Analysis in Agriculture: Current Trends and Recent Applications. Springer International Publishing (2018)
37. Quinn, B., Schiel, K., Caruso, G.: Mapping uncertainty from multi-criteria analysis of land development suitability, the case of Howth. Dublin. J. Maps. **11**, 487–495 (2015)
38. Mosadeghi, R., Warnken, J., Tomlinson, R., Mirfenderesk, H.: Uncertainty analysis in the application of multi-criteria decision-making methods in Australian strategic environmental decisions. J. Environ. Plan. Manag. **56**, 1097–1124 (2013)
39. Hamsa, K.R., Veerabhadrappa, B.: Review on decision-making under risk and uncertainty in agriculture. Econ. Aff. **62**, 447–453 (2017)
40. Cancian, F.: Risk and uncertainty in agricultural decision making. Risk Uncertain. Agric. Decis. Mak., 161–176 (1980). Ch. (7)
41. Klir, G.J., Folger, T.A.: Fuzzy Sets, Uncertainty and Information. Prentice Hall, Englewood Cliffs (1988)
42. Durbach, I.N., Stewart, T.J.: Modeling uncertainty in multi-criteria decision analysis. Eur. J. Oper. Res. **223**, 1–14 (2012)
43. Skalna, I., Rębiasz, B., Gaweł, B., Basiura, B., Duda, J., Opiła, J., Pełech-Pilichowsk, T.: Advances in Fuzzy Decision Making - Theory and Practice. Springer International Publishing (2015)
44. Al-Kloub, B., Al-Shemmeri, T., Pearman, A.: The role of weights in multi-criteria decision aid, and the ranking of water projects in Jordan. Eur. J. Oper. Res. **99**, 278–288 (1997)
45. Zardari, N.H., Ahmed, K., Shirazi, S.M., Yusop, Z.B.: Weighting Methods and Their Effects on Multi-Criteria Decision Making Model Outcomes in Water Resources Management. Springer International Publishing (2015)
46. Hwang, C.-L., Yoon, K.: Multiple Attribute Decision Making: Methods and Applications A State-of-the-Art Survey. Springer, Heidelberg (1981)
47. Evans, G.W.: Multiple Criteria Decision Analysis for Industrial Engineering: Methodology and Applications. CRC Press, Boca Raton (2016)
48. Yuexin, Y.: Efficiency measurement of agricultural mechanization in China based on DEA-TOPSIS models. World Autom. Congr. **2012**, 1–4 (2012)
49. Yuexin, Y.: Evolution analysis of agricultural mechanization in Jilin province based on TOPSIS methodology. World Autom. Congr. **2012**, 1–4 (2012)
50. Budianto, A.E., Yunus, E.P.A.: Expert system to optimize the best goat selection using topsis: decision support system. In: 2017 4th International Conference on Computer Applications and Information Processing Technology (CAIPT), pp. 1–5 (2017)

51. Papathanasiou, J., Ploskas, N., Bournaris, T., Manos, B.: A decision support system for multiple criteria alternative ranking using TOPSIS and VIKOR: a case study on social sustainability in agriculture. In: Decision Support Systems VI - Addressing Sustainability and Societal Challenges, pp. 3–15. Springer, Cham (2016)
52. Xiao, C., Shao, D., Yang, F.: Improved TOPSIS method and its application on initial water rights allocation in the watershed. In: Information Computing and Applications, pp. 583–592. Springer, Heidelberg (2011)
53. Yal, G.P., Akgün, H.: Landfill site selection utilizing TOPSIS methodology and clay liner geotechnical characterization: a case study for Ankara. Turkey. Bull. Eng. Geol. Environ. **73**, 369–388 (2014)
54. Seyedmohammadi, J., Sarmadian, F., Jafarzadeh, A.A., Ghorbani, M.A., Shahbazi, F.: Application of SAW, TOPSIS and fuzzy TOPSIS models in cultivation priority planning for maize, rapeseed and soybean crops. Geoderma **310**, 178–190 (2018)
55. Tan, Y., Cai, Z., Qi, H.: A process-based performance analysis for closed-loop agriculture supply Chain. In: 2010 International Conference on Intelligent System Design and Engineering Application, pp. 145–149 (2010)

Smart Management System to Monitor the Negative Impact of Chemical Substances and the Climate Change on the Environment and the Quality of Agricultural Production

Loubna Cherrat[1(✉)], Maroi Tsouli Fathi[2], and Mostafa Ezziyyani[2]

[1] University Chouaib Doukkali, El Jadida, Morocco
cherratloubna2@gmail.com
[2] Université Abdelmalek Essaâdi, Tangier, Morocco
maroi.tsouli@gmail.com, ezziyyani@gmail.com

Abstract. This paper has a twofold objective: on the one hand it focuses mainly on the study of agricultural substances that lead to the pollution of the environment and have a direct negative impact on agricultural production. Knowing today there are about fifteen controlled substances. Other non-regulated environmental pollutants are the subject of our research for monitoring and quality control. Currently no threshold concentration limit for these unregulated products (in air, or in water) exists or / and adopted. Therefore, our objective is to initiate research to define a list of pollutants to be investigated before launching a harmonized measurement test campaign. Some of these techniques are now widely used in farms. To extend this work and in the most comprehensive approach possible, our research team is working on the development of farm-level practices to integrate different environmental concerns: water quality, air. In order for these changes in practice to be truly appropriate for farmers and therefore have an effect on air and water quality, they must be coherent, practical and agronomic, and economically viable. On the other hand, the aim of this project is to show the impact of climate change on the agricultural sector and on agricultural production, based on data analysis of all parameters that cause climate change in Coordination with the necessary agricultural conditions. This will allow us to set up a climate change prediction system for the determination of the parameters of a risk prediction model and the prediction of periods of drought and flood. Our final ambition for this project is the realization of in-depth scientific and experimental field research by specialists and experts in the field (Applied to the two different regions of EL Jadida and Larache), to set up a system that helps decision-makers and farmers protecting the environment and production and adapting crops according to climatic variations. This system consists of several layers that ensure proper functioning of the process from the collection of data in real time via networks of wireless sensors, geolocation and the transmission of data via radio waves (5th generation) for possible filtering, Cleaning, storage and analysis in order to define an incremental knowledge database that can be used in the decision support system. This system can be used as a reference and generalized at the international level.

M. Ezziyyani (Ed.): AI2SD 2018, AISC 911, pp. 118–127, 2019.
https://doi.org/10.1007/978-3-030-11878-5_13

Keywords: Environment · Agriculture · Air · Water · Pollution · Big Data · Data analytics · Decision support system · WSN · Knowledgebase · GIS

1 Description

1.1 Objectif

Environmental pollution is defined as the introduction by man, directly or indirectly, into the nature of substances that may negatively impact the environment or more particularly indirectly the health and quality of human life. Environmental pollution is a topical issue is a preliminary preoccupation of Ministry of Environment pre-established in the National Strategy of Sustainable Development SNDD. The different sources of emissions are the residential and tertiary sectors, transport, industry and the agricultural sector. Therefore, in this paper our research focuses mainly on agricultural substances lead to environmental pollution and the effect of climate change that have a negative impact directly on the quality of agricultural production [1–3].

Indeed; Today there are the unregulated substances polluting, and which negatively affect the quality of agricultural production are the subject of our research for the monitoring and control of its quality. Currently no threshold concentration limits or techniques of use on these unregulated products (in air, or in water) exist or / and adopted. Therefore, our objective is to initiate research to define a list of pollutants to be investigated before launching a harmonized measurement test campaign. Some of these techniques are now widely used in farms [4–6]. To extend this work and adapt it to our regions, and in the most comprehensive approach possible, our research team is working on the evolution of practices at the farm level to integrate the different environmental concerns: Water and air. In order for these changes in practice to be truly appropriate by farmers and therefore have an effect on air and water quality, they must be coherent, practically and economically feasible, and economically viable.

The field of agriculture is extremely sensitive to climate change, intra- and inter-seasonal variations lead to an increase in temperatures, which reduces the yields of seasonal crops. Changing rainfall patterns increases the likelihood of short-term crop failures and a decline in long-term production. It will be interesting to analyze the impact of climate change on agriculture and agricultural production in particular to determine whether climate change poses a significant threat to agricultural production. Climate change is likely to degrade irreversibly the stock of natural resources on which agriculture depends [7]. The relationship between climate change and agriculture is twofold: agriculture contributes in many respects to climate change, and climate change generally has a negative impact on agriculture. Indeed, the moderate increase in local temperatures can have an effect is most often negative on agricultural yields. Some negative effects are already being felt in many parts of the world; the continuation of global warming will have an increasing impact in all regions. Extreme weather events (floods and droughts) are increasing and their frequency and severity are likely to increase, with serious consequences for food production and food security in all regions.

Otherwise, the moderate increase in temperatures allows for favorable environments for the rapid multiplication of aggressive fungi and the invasion of insects with the appearance of new species. To deal with these, the use of chemical substances becomes an absolute necessity. These substances have a double negative effect on the quality of production and the pollution of the environment. Conversely, the use of certain chemicals (N03) aggravates the situation of the greenhouse effect [8].

As a result, these issues will require a review and monitoring of chemical use and weather forecasting to address the effects of extreme precipitation patterns, intra- and inter-seasonal variations, and increased rates Evapotranspiration in all types of ecosystems.

2 Originality and Innovative Aspect

The proposed system will generate innovation in several knowledge areas that will be transmitted to society in order to reach sustainability for the agriculture activity:

- The collaboration between experts from agriculture and environmental monitoring will generate a model of interaction between agriculture and the environment related to the pollution.
- The elaboration of a clear list of indicators useful for monitoring purposes.
- The collaboration between agriculture experts and experts on sensor development is needed to find out the most adequate sensors for monitoring different parameters.
- The creation of a trademark for Eco-Friendly Agriculture Products will increase the market value of the products of the fields that adopt the proposed technologies. The Eco-Friendly trademark will attract new consumers to the market.

The research developed under this project has the potential to generate innovations and breakthroughs in different areas. The main one is the creation of a smart decision systems based on the wireless sensor network, TV white space and data analytics. This innovation includes areas of Biocontrole, Biochemistry, Big Data, electronics, physics of environment, GIS and telecommunication among others. The beneficiaries of those innovations will be in first place the agriculturists that will have the technology for monitoring its fields and quality production. The secondary beneficiaries are the environment; the reduction of chemical substance usage due to the optimization will permit the maintenance of ecological environment flow. Finally, the last beneficiaries are the general society that can enjoy from sustainable agriculture and can have eco-friendly citric in the market [9, 10].

Our activities are based on the identification of the behavior of chemical substances and polluting sources of the environment with the exploitation of the basic history of meteorological data, statistical yearbooks and meteorological reports of certain region of Morocco, reports and data on Agro climatic indices for certain crops. Analyzing these indicators with dataming algorithms will help us to determinate the characteristics of climate change and their impact on agriculture. The use of new technologies such as data analysis based on datamining algorithms, and an invaluable resource for planning, forecasting and decision-making to limit the negative impact of climate change on agricultural production. Indeed, anticipating the impact of short- and long-term climatic changes on agricultural production will use data-based algorithms based on estimation with the expectation of climate-specific constraints for each crop in this way. Diversification and adaptation options in relation to agricultural production.

In order to achieve these objectives, our team is composed of complementary specialties from different domains involved in the themes involved in this Bio-Control project, Physico-Chemistry Analysis, Geographic Information System, Big Data Analyst, Decision Support System And Data Transmission.

3 Scientific and Economic Impact

The active promotion of the outcome of the project activities belong to the main objectives of the Consortium. It is our purpose to maximize its impact and interest in Morocco and there regions and at global levels. The impact that will be performed for exploitation of the project results are divided into the following ones:

1. The scientific and technical results of this project by publications and demonstrations at top journals, conferences and workshops. The partners are committed to technical publications in high-quality conferences and journals.
2. Organizing scientific events like workshops devoted to the topics of the project. The Consortium plans the organization of International Conference directed to the topic of Big Data, Wireless Sensor Networks, TV White Space, Biocontrol and environment. The partners are already involved in several workshops, conferences, and other scientific venues related with the topic of the project.
3. Publishing research books with the technical reports of the research works performed and publishing conference or workshop proceedings. There is some experience in the consortium editing books and conference proceedings.
4. Publishing the papers written from this consortium in international journals. This can be achieved organizing special issues or directly submitting the papers to the international journals.
5. One of the main explicit objectives of the action is the proposal of smart decision system for agriculture based on wired & wireless sensor networks, which that the operation of such sustainably changed.
6. Involvement of industrial companies will ensure the continuity of the present project. The companies will helps to develop the proposals in the future and to create an exploitation plan.

4 Identification of the Components

4.1 Related Work and Scientific Objectives

The challenges faced by agriculture are more than ever relevant: Improving the nutritional and gustatory quality of production, ensuring economic stability for farmers, reducing environmental pollution and diseases (fertilizers, pesticides), along with coping with climate change. Through this WP, the research axes that we will discuss illustrate the answers that the relationship between modeling tools, assimilation of data and the IIoT, can bring to these stakes when manipulating mass data (Big Data). The new observation tools (embedded sensors, remote sensing, GIS) and all the data obtained via IIoT allow the collection of a large amount of data to better understand the environment and the state of the plants in real time Their coupling, through assimilation and data analysis technologies, to mathematical models and statistics allows to better understand the behavior of plants and the environment with regard to polluting products and to deduce from them sustainable and clean tools for crop management assistance [11–13]. Outsourcing and sharing of these data make it possible to understand the origins of the problems and to improve agronomic practices, to optimize warning systems, to model plant growth, to prevent climatic hazards on crop conditions and to protect the environment. The accumulation of data will be increasingly valuable in the years to come, as we will need a careful and precise monitoring of the response of plants to the quality of the environment and to the warming climate in progress.

4.2 Problematic

The complexities of this WP resides in (1) extracting, inventorying and deploying real data in a Big Data environment, (2) combining the approach of increasing data storage capacities through solutions enabling a progressive adjustment of needs (3) development of databases and algorithms capable of continuously extracting relevant lessons and knowledge from an evolving mass of multi-structured data (4) adaptation of Big Data Analytics solutions in the context of Research project with use of NoSQL, Redis, Hadoop or HBase without any surprises (5) the implementation of these optimized computing solutions requiring to resort to experts in mass data management and analysis, Data analytics, Data officer …).

1. ***Global Expected results***

   - Monitoring of crops conditions
   - Identify stressed plants
   - Estimate water content by remote sensing
   - Monitoring the agricultural production
   - Estimation and prediction of agricultural yields
   - Crops Identification
   - Mapping of agricultural parcels
   - Set up a system for an Agriculture of Precision
   - Estimate the number of plants and future yield
   - Early detection of invasions of insects and mushrooms
   - Satellite data in the agriculture of precision
   - Use of GIS in the agriculture of precision

2. ***Proposed methodology***

   - Extraction/Analysis/Exploration agricultural data with large volume of information.
   - Use of information technologies and analysis of mass data.
   - Automation of the data extraction process to obtain interesting knowledge and regularities and forecasts in the medium and long term.
   - Application and development of datamining algorithms
   - Geolocation and study of land use via real-time specific sensors
   - Detection of crop risks, particularly in the areas concerned, in order to deal only when necessary, for a sustainable and efficient agriculture

3. ***Socio-economic Impact***

   - Fighting diseases and pests
   - Optimizing the production process
   - Ensuring quality of production
   - Identify fertilizer and pesticide requirements
   - Combating pollution
   - Use less chemicals

5 Deliverables List

Ref	Title
DL5.1	Implementation of an interoperable IoT software layer for collection, filtering and automating heterogeneous data integration
DL5.2	Modeling and Implementation of a Big Data solution, for distributed and optimized storage of data related to the environment, agriculture and climate change
DL5.3	Set up a customized and cooperative distributed analysis tool for volume data for the extraction and internationalization of knowledge
DL5.4	Implement a system of agricultural decision support and environmental protection based on datamining and heuristic mathematical models
DL5.5	Geomatic modeling by multicriteria evaluation for the development of agricultural fields and prospecting for the risk of environmental pollution

5.1 Description of Deliverables

Deliverable Reference	*DL5.1*
Deliverable title	Implementation of an interoperable IoT software layer for collection, filtering and automating heterogeneous data integration
Objectives of deliverable	This important Task will provide the collection and the aggregation of data from multi-agricultural data resources and their careers to provide a detailed knowledge of how, when, the quantity of the resources used to cultivate and irrigate crops and how well the current environment affect growth stages. The Quality, efficiency of data and open access are at the heart of this WP. At this step, we will collect both biological data of the plants and real-time data of their environment with the longer-term aim of making these data accessible to both Farmer and to the Decision-making system. As a solution, we refer to Big Data solutions to manage large and disparate volumes of data being created by Farmer, tools, sensors and machines
Expected results	Using Big Data technologies requires new, innovative and scalable technology to collect, host and analytically process the vast amount of data gathered in order to derive real-time business insights that relate to resources use for an agriculture of precision, for performance, and for productivity management and enhancement of shareholder value
Deliverable Reference	DL5.2
Deliverable title	Modeling and Implementation of a Big Data Analytics solution, for distributed and optimized storage of data related to the environment, agriculture and climate change
Objectives of deliverable	For monitoring the environment, agriculture and climate change, multiple actors of Internet technology are producing very large amounts of data. All generate real-time extending information based on the 3 Vs of Gartner: Volume, Velocity and Variety. In order to efficiently distribute this data and optimize the data storage related to the environment, agriculture and climate change, we will provide a new solution to keep track of the dynamic aspect of their chronological evolution by means of two main approaches: first, by a dynamic model able to support type changes every second with a successful processing and second, the support of data volatility using an intelligent model, taking into consideration key-data without processing all volumes of history and up to date data
Expected results	The Expected results of this study is to establish, based on these approaches, an integrative vision of data life cycle set on 3 steps, (1) data synthesis by selecting key-values of micro-data acquired by different data source operators, (2) data fusion by sorting and duplicating the selected key-values based on a normalization aspect in order to get a faster processing of data and (3) the data transformation into a specific format of maps, via Hadoop in the standard Map Reduce process, in order to create a customized processing chain of Big Data

(*continued*)

(*continued*)

Deliverable Reference	*DL5.3*
Deliverable title	Set up a customized and cooperative distributed analysis tool for volume data for the extraction and internationalization of knowledge
Objectives of deliverable	This task focusing upon methodologies for extracting useful knowledge from data and classification for research optimization. To make the prepared data set useful for demand forecasting and pattern extraction, we pre-process the data set using a novel approach based on data mining knowledge. In this context, the analytic data that has been collected over time is used to make processes more efficient, effective, predictable, and profitable for increasing the food production by enhancing the efficient use of products
Expected results	setting up a spatialized knowledge base to serve as an information reference for the multicriteria construction of a suitability map indicating high potential areas for cultivation
Deliverable Reference	*DL5.4*
Deliverable title	Implement a system for agricultural decision support and environmental protection based on datamining and heuristic mathematical models
Objectives of deliverable	In this d*eliverable,* we would like to apply a decision approach technique to forecast future need. We also develop a web based decision support system for the managers, farmers and researchers in order to access various data including the prediction of possible agriculture requirement in future. We evaluate the pre-processing techniques based on substances use and food production quality. The challenge of performing the decision support from extracting knowledge draws upon research in statistics, databases, pattern recognition, machine learning, data visualization, optimization, and high-performance computing, to deliver advanced business intelligence and web discovery solutions
Expected results	Decision-making tool to better manage the environment, spending, quality of production and crop orientation
Deliverable title	Geomatic modeling by multicriteria evaluation for the development of agricultural fields and prospecting for the risk of environmental pollution
Objectives of deliverable	The purpose of the project is to prospect by geomatic method the ability of the agricultural soil to receive a type of agriculture in a prosperous and sustainable way for its consideration in the scheme of coherence in specific region with relation to its environment. One of the most promising aspects of geographic information systems (GIS) is their ability to contribute to decision-making by spatial and temporal modeling. Multicriteria assessment (environmental, biophysical, behavioral, climate change) is particularly useful for the location of areas suitable for use. It is based on the assumption that for a given date there exists a series of spatialized criteria that can explain the variability of states and the need for land use and for a specific crop

(*continued*)

(*continued*)

Expected results	– Constitute a pedological database from which several spatial agricultural maps could be georeferenced – Adjustment of cultural practices to the heterogeneous characteristics of the plot (tillage, irrigation, sowing, pollution) – Accurately target areas to be treated or grown – Improved management of input costs (fertilizers, seeds, fungicides, herbicides, etc.) – Limiting leaching of excess fertilizer to groundwater – Identification of spatial constraints of agriculture in different regions – Analysis of spatial temporal factors for the identification (1) of agricultural sites, (2) up-to-date land use, (3) soil morpho soil characterization

References

1. Ezziyyani, M., Hamdache, A., Barka, N., Sebbar, H., Mouraziq, F., Lamarti, A., Requena, A.M., Egea-Gilabert, C., Requena, M.E., Candela Castillo, M.E.: Variations de la répartition géographique, taille et teneur en cadmium chez la moule Mytilus galloprovincialis prélevée au niveau du littoral de Safi. J. Mater. Environ. Sci. **5**(S2), 2459–2466 (2014)
2. Ezziyyani, M., Pérez-Sánchez, C., Sid Ahmed, A., Requena, M.E., Candela, M.E.: Trichoderma harzianum as biofungicida in the control of Phytophthora capsici in pepper (Capsicum annuum L.) plants. Anales de Biología **26**, 35–45 (2004)
3. Ezziyyani, M., Hamdache, A., Requena, A., Egea-Gilabert, C., Candela, M.E., Gonzalez, R. L., Requena, M.E.: Improving the antioomycete capacity in vivo and in vitro of a formula containing a combination of antagonists of Phytophthora capsici Leonian. Anales de Biología **33**, 67–77 (2011)
4. Ezziyyani, M., Sid Ahmed, A., Pérez-Sánchez, C., Requena, M.E., Candela, M.E.: Biocological control by antagonistic microorganisms. Example in the interaction pepper (Capsicum annuum L.) - Phytophthora capsici cause of the sadness. Horticom News. **191**, 8–15 (2006)
5. Ezziyyani, M., Requena, M.E., Candela Castillo, M.E.: Production of PR-proteins during the induction of resistance to Phytophthora capsici in pepper (Capsicum annuum L.) plants treated with Trichoderma harzianum. Anales de Biología. **27**, 173–180 (2005)
6. Ezziyyani, M., Requena, M.E., Pérez-Sánchez, C., Candela, M.E.: Effect of substrate and temperature on the biological control of Phytophthora capsici in pepper (Capsicum annuum L.). Anales de Biología **27**, 473–480 (2005)
7. Ezziyyani, M., Requena, M.E., Lamarti, A., Candela Castillo, M.E., Pérez-Sánchez, C., Requena, M.E., Rubio, L., Candela, M.E.: Biocontrol of pepper (*Capsicum annuum*) root rot caused by *Phytophthora capsici* using *Streptomyces rochei* Ziyani. Anales de Biología. **26**, 69–78 (2004)
8. Sadiq, A.S., Almohammad, T.Z., Khadri, R.A.B.M., Ahmed, A.A., Lloret, J.: An energy-efficient cross-layer approach for cloud wireless green communications. In: IEEE Second International Conference on Fog and Mobile Edge Computing (FMEC), pp. 230–234, May 2003
9. Mehmood, A., Lloret, J., Sendra, S.: A secure and low-energy zone-based wireless sensor networks routing protocol for pollution monitoring. Wirel. Commun. Mobile Comput. **16** (17), 2869–2883 (2016)

10. Mehmood, A., Khan, S., Shams, B., Lloret, J.: Energy-efficient multi-level and distance-aware clustering mechanism for WSNs. Int. J. Commun Syst **28**(5), 972–989 (2015)
11. Parra, L., Karampelas, E., Sendra, S., Lloret, J., Rodrigues, J.J.: Design and deployment of a smart system for data gathering in estuaries using wireless sensor networks. In: IEEE International Conference on Computer, Information and Telecommunication Systems (CITS), pp. 1–5, July 2015
12. Lloret, J., Garcia, M., Bri, D., Sendra, S.: A wireless sensor network deployment for rural and forest fire detection and verification. Sensors **9**(11), 8722–8747 (2009)
13. Bri, D., Garcia, M., Lloret, J., Dini, P.: Real deployments of wireless sensor networks. In: IEEE Third International Conference on Sensor Technologies and Applications. SENSORCOMM 2009, pp. 415–423, June 2009

Author Index

M. Ezziyyani (Ed.): AI2SD 2018, AISC 911, pp. 129–130, 2019.
https://doi.org/10.1007/978-3-030-11878-5

Z

Printed in the United States
By Bookmasters